OXFORD MEDICAL PUBLICATIONS

Critical Care Algorithms

Critical Care Algorithms

R. F. Armstrong
Consultant Anaesthetist,
ICU, University College Hospital, London

Catherine Bullen
Consultant Anaesthetist, ICU,
Middlesex Hospital, London

S. L. Cohen
Consultant Physician, ICU,
University College Hospital, London

Mervyn Singer
Lecturer in Intensive Care
and General Medicine,
University College and Middlesex
Hospital Medical School, London

A. R. Webb
Consultant Physician, ICU,
Middlesex Hospital, London

OXFORD   NEW YORK   TOKYO
OXFORD UNIVERSITY PRESS
1991

Oxford University Press, Walton Street, Oxford OX2 6DP

Oxford   New York   Toronto
Delhi   Bombay   Calcutta   Madras   Karachi
Petaling Jaya   Singapore   Hong Kong   Tokyo
Nairobi   Dar es Salaam   Cape Town
Melbourne   Auckland
and associated companies in
Berlin   Ibadan

Oxford is a trade mark of Oxford University Press

Published in the United States
by Oxford University Press, New York

A catalogue record for this book is available from the British Library

Library of Congress Cataloging in Publication Data
Intensive care algorithms / R. F. Armstrong ... [et al.].
(Oxford medical publications)
Includes bibliographical references.
1. Critical care medicine.   2. Medical protocols.   I. Armstrong,
R. F. (Rod F.)   II. Series.
[DNLM: 1. Algorithms.   2. Critical Care—methods.   WX 218 I60417]
RC86.7.I54   1991   616'.028–dc20   91–27029
ISBN 0–19–262107–6 (pbk.)

Set by
Footnote Graphics, Warminster, Wiltshire
Printed in Great Britain
by Bookcraft (Bath) Ltd
Midsomer Norton, Avon

Preface

There are many textbooks in the field of intensive care although few
are devoted primarily to patient management. With this book we
have documented our approach to 36 procedures and situations
commonly seen in intensive care medicine. Necessary justifications
and amplification are given as short notes and references following
each algorithm.

These algorithms are not intended as a substitute for detailed
clinical assessment and thought. They are not necessarily the only
way of managing a particular problem; they are simply our approach
which has been tried and tested. We would strongly encourage
individual units and doctors to read and amend them as they see fit
for their particular practice or patient. If they can vindicate their
changes then the problem has been truly considered!

It should also be remembered that every treatment step carries the
risk of associated side effects, particularly the invasive procedures.
The possibility of iatrogenic injury must always be borne in mind yet
it is not possible to cover every eventuality within manageable
algorithms. The art of medicine lies in the balance of risk and we
hope that this book will help in achieving this balance.

We would like to thank the staff of the Middlesex and University
College Hospitals for their assistance and helpful criticism. In
particular we would like to thank Sister Sheila Adam, Staff Nurse
Jake Jacobsen, Charge Nurse Bernard Place, Senior Dietitian Alexa
Scott, Pharmacist Elma Pryce-Jones, Mr Peter Goldstraw, Mr Chris
Rudge and Drs Howard Swanton, Adrian Thrasher, Roland Valori
and Malcolm Walker.

December 1990
R.A.
C.B.
S.C.
M.S.
A.W.

Contents

Abbreviations

Respiratory system (RS)

ARDS	adult respiratory distress syndrome
CPAP	continuous positive airways pressure
CXR	chest X-Ray
ET	endotracheal
F	frequency
FiO_2	inspired oxygen concentration
FRC	functional residual capacity
I : E ratio	inspiratory : expiratory ratio
IMV	intermittent mandatory ventilation
IPPV	intermittent positive pressure ventilation
IPS	inspiratory pressure support
MMV	mandatory minute ventilation
PE	pulmonary embolus
PEEP	positive end-expiratory pressure
PEFR	peak expiratory flow rate
Qs/Qt	intrapulmonary shunt
SIMV	synchronized intermittent mandatory ventilation
V_d	physiological dead space
V_E	minute ventilation
V_t	tidal volume
V/Q	ventilation/perfusion

Neurological

CFAM	cerebral function analysis monitor
CPP	cerebral perfusion pressure
CSF	cerebrospinal fluid
GCS	Glasgow Coma Score
ICP	intracranial pressure

Renal

ADH	antidiuretic hormone
ATN	acute tubular necrosis
CSU	catheter specimen of urine
DDAVP	arginine vasopressin
TUR(P)	transurethral resection (prostate)

Haematological

ACT	activated clotting time
APTT	activated partial thromboplastin time
COHb	carboxyhaemoglobin
DIC	disseminated intravascular coagulation
FBC	full blood count

FFP	fresh frozen plasma
WBC	white blood count

Cardiovascular system

AF	atrial fibrillation
AP	arterial pressure
AV	atrioventricular
CI	cardiac index
CO	cardiac output
CPK	creatine phosphokinase
CPR	cardiopulmonary resuscitation
CVP	central venous pressure
DO_2	oxygen delivery
DVT	deep vein thrombosis
GTN	glyceryl trinitrate
HR	heart rate
LV	left ventricle
LVEDP	left ventricular end-diastolic pressure
LVF	left ventricular failure
MAP	mean arterial pressure
MI	myocardial infarction
PA	pulmonary artery
PA_{dias}	diastolic pulmonary artery pressure
PA_{sys}	systolic pulmonary artery pressure
PAWP	pulmonary artery wedge pressure
RA	right atrium
RV	right ventricle
SaO_2	arterial oxygen saturation
SvO_2	mixed venous oxygen saturation
SVR	systemic vascular resistance
SVT	supraventricular tachycardia
VF	ventricular fibrillation
VO_2	oxygen consumption
VSD	ventricular septal defect
VT	ventricular tachycardia

Miscellaneous

bd	twice a day
BMR	basal metabolic rate
CHO	carbohydrate
DKA	diabetic ketoacidosis
ECF	extracellular fluid
GM $-$ve	Gram negative
HB_sAg	Hepatitis B surface antigen
HIV	human immunodeficiency virus
I/V	intravenous
LFTs	liver function tests
M,C & S	microscopy, culture & sensitivity
MAST	medical anti-shock trouser

MCTs	medium chain triglycerides
NG	nasogastric
NSAID	non-steroidal anti-inflammatory drug
P/O	per oral
P/R	per rectum
Rx	treatment
S/C	subcutaneous
S/L	sublingual
tds	three times a day
TPN	total parenteral nutrition
U&E	urea and electrolytes

Formulae and normal ranges

Formulae for derived variables	Units	Normal
Systemic vascular resistance		
$SVR = \dfrac{MAP - RAP}{CO} \times 79.9$	$dyn.s.cm^{-5}$	960–1400
Pulmonary vascular resistance		
$PVR = \dfrac{MPAP - PAWP}{CO} \times 79.9$	$dyn.s.cm^{-5}$	25–125
Left ventricular stroke work		
$LVSW = (MAP - PAWP) \times SV \times 0.0136$	$g\text{-}m.m^2.beat^{-1}$	44–68
Right ventricular stroke work		
$RVSW = (MPAP - RAP) \times SV \times 0.0136$	$g\text{-}m.m^2.beat^{-1}$	4–8
Arterial oxygen transport		
$DO_2 = 0.134 \times CO \times [Hb_a] \times SaO_2$	$ml.min^{-1}$	950–1300
Systemic oxygen consumption		
$VO_2 = 0.134 \times CO \times ([Hb_a] \times SaO_2 - [HB_v] \times SvO_2)$	$ml.min^{-1}$	180–320
Alveolar arterial oxygen gradient		
$(A\text{-}a)DO_2 = FiO_2 \times 94.8 - PaCO_2 - PaO_2$	kPa	<3
Intrapulmonary shunt		
$\dfrac{Qs}{Qt} = \dfrac{CcO_2 - CaO_2}{CcO_2 - CvO_2} \times 100$ where $CaO_2 = 0.0134 \times [Hb_a] \times SaO_2,$ $CvO_2 = 0.0134 \times [Hb_v] \times SvO_2$ $CcO_2 \approx 1.34 \times [Hb_a]\ ml.dl^{-1}$	%	<15

KEY TO ALGORITHMS

▭	Action	▬	Refer to other protocol
⬭	Observation	⚠	Important note
– ➤	Always refer back		

Critical Care Algorithms

1. General

1.1/1.2: Adult central venous catheter insertion/pulmonary artery catheter insertion

West's zones
1. Catheter tip should lie in a zone III region.
2. The catheter tip should lie below the level of the left atrium on a lateral CXR.
3. Suspect a non-zone III position if:
   - Rise in wedge pressure greater than half any PEEP increment.
   - No detectable cardiac pulsation.
   - Excessive respiratory variation in the wedge trace.
4. A non-zone III position is more likely with PEEP and/or hypovolaemia.

Timing of measurements
1. Wedge pressure readings are taken at end-expiration. At this point intra-thoracic pressure approximates closest to atmospheric pressure.
   - For ventilated patients and expiration $\equiv$ lowest wedge reading.
   - For spontaneous breathing and expiration $\equiv$ highest wedge reading.
   - During spontaneous breathing in the dyspnoeic patient, use the 'mean' wedge reading.
2. The lowest wedge reading is equivalent to the diastolic and the highest equivalent to the systolic wedge reading on most bedside monitors.
3. The 'diastolic' reading may represent a spontaneous inspiration during IMV. A print-out of the trace should be scrutinized.
4. CVP, wedge pressure and cardiac output should not be measured during rapid volume infusion.
5. Cardiac output measurements should be spread throughout the respiratory cycle.

Valve lesions
1. Cardiac output measurements are unreliable in tricuspid regurgitation.
2. Wedge pressure $\neq$ LVEDP in mitral stenosis.
3. Read wedge pressure on the end of the 'a' wave in mitral regurgitation.

Infection
1. Consider replacing PA catheter at 72–96 hourly intervals and central venous catheters after 6 days.

Bibliography
Ciggaroa RG, Lange RA, Williams RH, Bedotto JB, and Hillis LD. Under-estimation of cardiac output by thermodilution in patients with tricuspid regurgitation. Am J Med. 1989; 86: 417–420.

Gill RT, Kruse JA, Thill-Baharozian MC, and Carlson RW. Triple vs Single lumen central venous catheters. A prospective study in a critically ill population. Arch Intern Med. 1989; 149: 1139–1143.

Jansen JRC and Versprille A. Improvement of cardiac output estimation by the thermodilution method during mechanical ventilation. Intensive Care Med. 1986; 12: 71–79.

Matthay MA and Chatterjee K. Bedside catheterisation of the pulmonary artery: Risks compared with benefits. Ann Intern Med. 1988; 109: 826–834.

O'Quin R and Marini JJ. Pulmonary artery occlusion pressure: Clinical physiology, measurement and interpretation. Am Rev Respir Dis. 1983; 128: 319–326.

Rice DL, Awe RJ, Gaaseh WH, Alexander JK, and Jenkins DE. Wedge pressure measurement in obstructive pulmonary disease. Chest. 1974; 66: 628–632.

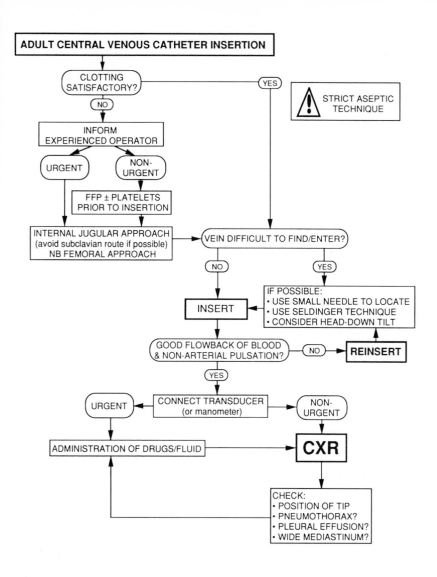

ADULT CENTRAL VENOUS CATHETER INSERTION

CLOTTING SATISFACTORY?

NO

YES

⚠ STRICT ASEPTIC TECHNIQUE

INFORM EXPERIENCED OPERATOR

URGENT

NON-URGENT

FFP ± PLATELETS PRIOR TO INSERTION

INTERNAL JUGULAR APPROACH
(avoid subclavian route if possible)
NB FEMORAL APPROACH

VEIN DIFFICULT TO FIND/ENTER?

NO

YES

INSERT

IF POSSIBLE:
• USE SMALL NEEDLE TO LOCATE
• USE SELDINGER TECHNIQUE
• CONSIDER HEAD-DOWN TILT

GOOD FLOWBACK OF BLOOD & NON-ARTERIAL PULSATION?

NO

REINSERT

YES

URGENT

CONNECT TRANSDUCER (or manometer)

NON-URGENT

ADMINISTRATION OF DRUGS/FLUID

CXR

CHECK:
• POSITION OF TIP
• PNEUMOTHORAX?
• PLEURAL EFFUSION?
• WIDE MEDIASTINUM?

West JB, Dollery CT, and Naimark A. Distribution of blood flow in isolated lung: relation to vascular and alveolar pressures. J Appl Physiol. 1964; 19: 713–724.
Wetzel RC and Latson TW. Major errors in thermodilution cardiac output measurement during rapid volume infusion. Anesthesiology 1985; 62: 684–687.

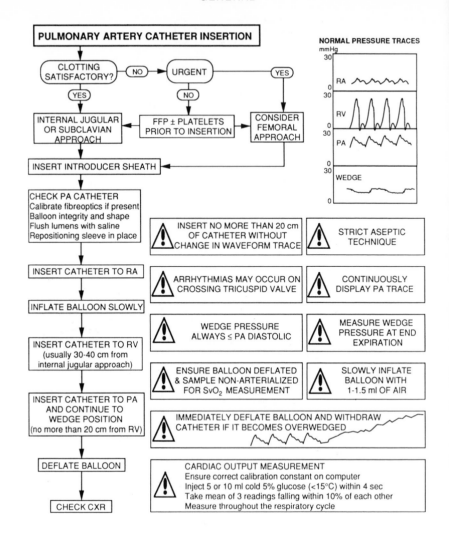

PULMONARY ARTERY CATHETER INSERTION

CLOTTING SATISFACTORY? —NO→ URGENT —YES→

YES ↓   NO ↓

INTERNAL JUGULAR OR SUBCLAVIAN APPROACH ← FFP ± PLATELETS PRIOR TO INSERTION → CONSIDER FEMORAL APPROACH

INSERT INTRODUCER SHEATH

CHECK PA CATHETER
Calibrate fibreoptics if present
Balloon integrity and shape
Flush lumens with saline
Repositioning sleeve in place

INSERT CATHETER TO RA

INFLATE BALLOON SLOWLY

INSERT CATHETER TO RV
(usually 30-40 cm from internal jugular approach)

INSERT CATHETER TO PA AND CONTINUE TO WEDGE POSITION
(no more than 20 cm from RV)

DEFLATE BALLOON

CHECK CXR

NORMAL PRESSURE TRACES
mmHg

RA
RV
PA
WEDGE

INSERT NO MORE THAN 20 cm OF CATHETER WITHOUT CHANGE IN WAVEFORM TRACE

STRICT ASEPTIC TECHNIQUE

ARRHYTHMIAS MAY OCCUR ON CROSSING TRICUSPID VALVE

CONTINUOUSLY DISPLAY PA TRACE

WEDGE PRESSURE ALWAYS ≤ PA DIASTOLIC

MEASURE WEDGE PRESSURE AT END EXPIRATION

ENSURE BALLOON DEFLATED & SAMPLE NON-ARTERIALIZED FOR SvO_2 MEASUREMENT

SLOWLY INFLATE BALLOON WITH 1-1.5 ml OF AIR

IMMEDIATELY DEFLATE BALLOON AND WITHDRAW CATHETER IF IT BECOMES OVERWEDGED

CARDIAC OUTPUT MEASUREMENT
Ensure correct calibration constant on computer
Inject 5 or 10 ml cold 5% glucose (<15°C) within 4 sec
Take mean of 3 readings falling within 10% of each other
Measure throughout the respiratory cycle

Notes

1.3: Fluid challenge

1. Hypovolaemia must be treated urgently to avoid sequelae such as organ failure.
2. Maintain a high index of suspicion for hypovolaemia.
3. Ensure an adequate circulating volume before considering other methods of circulatory support.
4. Repetitive infusion of small aliquots of colloid (or blood) is a safe and convenient method of managing hypovolaemia.
5. Critically ill patients require a greater than normal blood volume to maintain adequate circulatory function.

Central venous pressure and stroke volume responses
1. Peripheral vasoconstriction may maintain CVP and PAWP despite hypovolaemia.
2. The absolute CVP or PAWP is not a good guide to volume status particularly in chronic cardiac or pulmonary vascular diseases.
3. The response of CVP or PAWP to fluid is variable and must be measured between aliquots.
4. Stroke volume should be used rather than cardiac output because:
   ● Stroke volume rises if cardiac output falls less than heart rate.
   ● Stroke volume rises if cardiac output is constant but heart rate falls.
5. A reduction in stroke volume with fluid may represent subendocardial ischaemia or functional mitral regurgitation. No benefit will be gained by further volume infusion.

Which fluid?
1. Packed red blood cells have a high haematocrit. Additional fluid may be required for a volume response.
2. Crystalloid fluids are not an efficient means of expanding the plasma volume. Approximately ¾ of a crystalloid infusion is quickly lost from the plasma.
3. We currently recommend gelatin solutions for short term plasma expansion in simple hypovolaemia and hydroxyethyl starch when there is capillary leak.

Bibliography
Haupt MT and Rackow EC. Colloid osmotic pressure and fluid resuscitation with hetastarch, albumin, and saline solutions. Crit Care Med 1982; 10: 159–162.
Lazrove S, Waxman K, Shippy C, and Shoemaker WC. Hemodynamic, blood volume and oxygen transport responses to albumin and hydroxyethyl starch infusions in critically ill postoperative patients. Crit Care Med 1980; 8: 302–306.
Lundsgaard Hansen P and Pappova E. Colloids versus crystalloids as volume substitutes: clinical relevance of the serum oncotic pressure. Ann Clin Res 1981; 133: 5–17.
Manny J, Grindlinger GA, Dennis RC, Weisel RD, and Hechtman HB. Myocardial performance curves as a guide to volume therapy. Surg Gynecol Obstet 1979; 149: 863–873.
Shoemaker WC, Appel PL, Kram HB, Waxman K, and Lee TS. Prospective trial of supranormal values of survivors as therapeutic goals in high-risk surgical patients. Chest 1988; 94: 1176–1186.
Sturm JA and Wisner DH. Fluid resuscitation of hypovolemia. Intensive Care Med 1985; 11: 227–230.
Webb AR, Barclay SA, and Bennett ED. In vitro colloid osmotic pressure of commonly used plasma substitutes: a study of the diffusibility of colloid molecules. Intensive Care Med 1989; 15: 116–120.

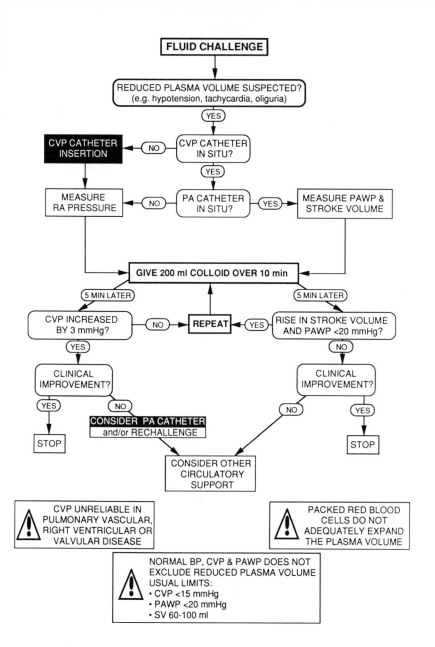

FLUID CHALLENGE

REDUCED PLASMA VOLUME SUSPECTED?
(e.g. hypotension, tachycardia, oliguria)

YES

CVP CATHETER
IN SITU?

NO → **CVP CATHETER INSERTION**

YES

PA CATHETER
IN SITU?

NO → MEASURE RA PRESSURE

YES → MEASURE PAWP & STROKE VOLUME

GIVE 200 ml COLLOID OVER 10 min

5 MIN LATER

CVP INCREASED BY 3 mmHg?

NO ← **REPEAT** → YES

RISE IN STROKE VOLUME AND PAWP <20 mmHg?

5 MIN LATER

YES

CLINICAL IMPROVEMENT?

NO

CLINICAL IMPROVEMENT?

YES → STOP

NO → **CONSIDER PA CATHETER** and/or RECHALLENGE

NO

YES → STOP

CONSIDER OTHER CIRCULATORY SUPPORT

⚠ CVP UNRELIABLE IN PULMONARY VASCULAR, RIGHT VENTRICULAR OR VALVULAR DISEASE

⚠ PACKED RED BLOOD CELLS DO NOT ADEQUATELY EXPAND THE PLASMA VOLUME

⚠ NORMAL BP, CVP & PAWP DOES NOT EXCLUDE REDUCED PLASMA VOLUME
USUAL LIMITS:
• CVP <15 mmHg
• PAWP <20 mmHg
• SV 60-100 ml

1.4: Sedation, analgesia, and paralysis

Sedation
1. There is a wide and unpredictable variation in patient response to drug infusions, particularly in the critically ill, due to alterations in hepatic and renal blood flow and function, in plasma protein levels, in distribution volumes, and in tolerance.
2. Regular re-assessment is mandatory if overdose is to be avoided.
3. Sedation scoring should be performed at regular intervals with subsequent alterations in drug dosage.

Reversal of sedation and analgesia
1. Great care should be exercised when reversing sedation.
2. Flumazenil (0.1 mg I/V increments or infusion of 0.5–1 mg/hr) can produce sudden awakening with biting of ET tube, anxiety and fear. Increases in intracranial pressure and convulsions have been reported. Do not use after head injury or epilepsy.
3. Failure of flumazenil to reverse the action of midazolam has been reported. It is short-acting with an elimination half life < 1 hr. Thereafter sedation may recur.
4. Naloxone is a competitive antagonist of morphine and other opioids. Duration of effect is 30 min. Large doses reverse analgesia; average dose given 200–400 μ I/V. Well tolerated with minimal side effects.

Paralysis
1. The paralysed patient should be adequately sedated.
2. Fentanyl and alfentanil may not produce adequate sedation/sleep unless used in very high dosage.
3. Because of persistence of muscle relaxants after continuous infusion a daily assessment of neuromuscular blockade is important using a train-of-four stimulation (2 Hz at 0.5 sec intervals). No response suggests 100% blockade and indicates dose reduction or cessation. Two twitches usually signifies adequate relaxation for ventilatory purposes. Use of a Datex Relaxograph facilitates this and allows more accurate measurement of twitch size and $T_4:T_1$ ratio.

Administration
1. Rapid bolus doses of sedatives/analgesics/anaesthetics in the critically ill should be discouraged. Slow injection with observation of heart rate and blood pressure should prevent unwanted hypotension. Small doses can be administered more rapidly.

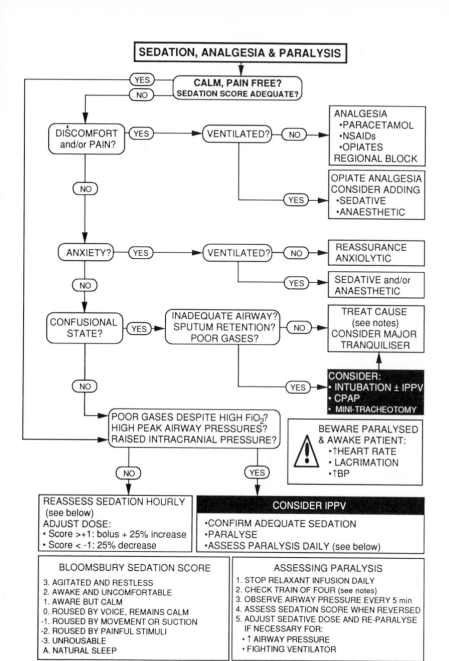

SEDATION, ANALGESIA & PARALYSIS

CALM, PAIN FREE?
SEDATION SCORE ADEQUATE? — YES / NO

DISCOMFORT and/or PAIN? — YES → **VENTILATED?**
- NO → ANALGESIA
  - •PARACETAMOL
  - •NSAIDs
  - •OPIATES
  - REGIONAL BLOCK
- YES → OPIATE ANALGESIA CONSIDER ADDING
  - •SEDATIVE
  - •ANAESTHETIC

(DISCOMFORT and/or PAIN?) — NO ↓

ANXIETY? — YES → **VENTILATED?**
- NO → REASSURANCE ANXIOLYTIC
- YES → SEDATIVE and/or ANAESTHETIC

(ANXIETY?) — NO ↓

CONFUSIONAL STATE? — YES → **INADEQUATE AIRWAY? SPUTUM RETENTION? POOR GASES?**
- NO → TREAT CAUSE (see notes) CONSIDER MAJOR TRANQUILISER
- YES → CONSIDER:
  - • INTUBATION ± IPPV
  - • CPAP
  - • MINI-TRACHEOTOMY

(CONFUSIONAL STATE?) — NO ↓

POOR GASES DESPITE HIGH FiO₂? HIGH PEAK AIRWAY PRESSURES? RAISED INTRACRANIAL PRESSURE?

⚠ BEWARE PARALYSED & AWAKE PATIENT:
- •↑HEART RATE
- • LACRIMATION
- •↑BP

- NO → **REASSESS SEDATION HOURLY** (see below)
  ADJUST DOSE:
  - • Score >+1: bolus + 25% increase
  - • Score < -1: 25% decrease
- YES → **CONSIDER IPPV**
  - •CONFIRM ADEQUATE SEDATION
  - •PARALYSE
  - •ASSESS PARALYSIS DAILY (see below)

BLOOMSBURY SEDATION SCORE

3. AGITATED AND RESTLESS
2. AWAKE AND UNCOMFORTABLE
1. AWARE BUT CALM
0. ROUSED BY VOICE, REMAINS CALM
-1. ROUSED BY MOVEMENT OR SUCTION
-2. ROUSED BY PAINFUL STIMULI
-3. UNROUSABLE
A. NATURAL SLEEP

ASSESSING PARALYSIS

1. STOP RELAXANT INFUSION DAILY
2. CHECK TRAIN OF FOUR (see notes)
3. OBSERVE AIRWAY PRESSURE EVERY 5 min
4. ASSESS SEDATION SCORE WHEN REVERSED
5. ADJUST SEDATIVE DOSE AND RE-PARALYSE IF NECESSARY FOR:
   • ↑ AIRWAY PRESSURE
   • FIGHTING VENTILATOR

DRUGS USED IN SEDATION, ANALGESIA & PARALYSIS

Drug	Bolus	Infusion	Duration after single dose	CVS	Resp'y system	Relevant ICU effects & problems
Morphine cost: cheap <£1/hr	0.1-0.2 mg/kg	0.05- 0.07 mg/kg/hr	· Elimination half-life 1.5 - 4 hr Longer after infusion in renal or hepatic impairment · Slow onset	· Occ bradycardia · Fall in SVR from histamine release · Mild fall in BP	Depression	· Analgesic & sedative · Occasional bronchospasm · Reduces gut motility and gastric emptying · Constricts sphincter of Oddi, raises CBD pressure · Active metabolites cause prolonged effects in renal failure (morphine-6-glucuronide) · ? impairs immune response
Pethidine cost: cheap <£1/hr	0.5 mg/kg	0.1-0.3 mg/kg/hr	· EHL approx 4 hr Slower after infusion and with hepatic dysfunction	· Fall in SVR · Mild fall in BP	Depression	· Analgesic & sedative · With continuous infusion norpethidine accumulates, causing fits & excitation · Reduces gut motility, raises CBD pressure, reduces gastric emptying
Fentanyl cost: medium <£5/hr	5-7.5 μg/kg	5-20 μg/kg/hr	· EHL approx 4 hr Much slower after prolonged infusion	· Minimal effect	Intense depression	· Analgesic · Poor sedative · Reduces GI motility · Muscle rigidity at high dose · Tendency to accumulate after continuous infusion
Alfentanil cost: dear <£10/hr	15-30 μg/kg	20-120 μg/kg/hr	· EHL 90 min May be slower after infusion	· Minimal effect	Intense depression	· Analgesic · Poor sedative · Reduces GI motility · CBD pressure rises
Midazolam cost: cheap <£1/hr	50 μg/kg slowly	50-100 μg/kg/hr	· EHL 2-4 hr Slower after infusion	· SVR falls · HR increases · Moderate fall in BP	Can cause respiratory depression -especially in elderly	· Sedative · Effects prolonged in hepatic & renal failure, after enzyme inhibition (cimetidine), and in elderly. · 10% are slow metabolisers.
Diazepam cost: cheap <£1/hr	0.15-0.2 mg/kg slowly	Intermittent injection preferred	· EHL 20-90 hrs	· Minimal effects	Can cause respiratory depression -especially in elderly	· Has active metabolites · Prolonged effects in hepatic & renal dysfunction, elderly. · Potentiates other CNS depressants. · Is potentiated by cimetidine
Propofol cost: medium <£5/hr	1-2 mg/kg slowly	1-3 mg/kg/hr	· Short-acting (30-60 mins) · No accumulation (EHL after infusion 3-6 hrs)	· SVR falls · Negatively inotropic · Can cause large fall in BP, esp. in poor CVS function or hypovolaemia	Can cause respiratory depression	· Anaesthetic · Pharmacokinetics not affected by renal or hepatic disease. · ? cleared more slowly in renal insufficency, elderly · ? reduces ICP, CPP · seizures reported · green urine caused by phenols · made up in 10% soya bean oil in water
Isoflurane cost: very expensive <£15/hr		Inhalation of 0.1-0.6% in air-oxygen mix	· Rapid recovery · Rapid response to change in conc'n	· Minimal · Some SVR fall · Mild fall in BP · May inhibit pulmonary hypoxic vasoconstriction	Minimal. Mild resp'y depression at higher conc'ns	· Anaesthetic · Minimal metabolism (0.2%) independent of liver/kidney. · Nephrotoxic level of fluoride reported after 5 days' use · Controversial effect on ICP. Seizures reported.
Chlor-methiazole cost: cheap <£1/hr	0.1-0.2 ml/kg/min	0.5-1.0 ml/min	· EHL about 8 hr · May be prolonged after continuous infusion	· Stable CVS · HR rises	Some resp'y depression	· Sedative, · Anticonvulsant · Metabolized in liver · Bronchorrhoea · Water overload possible · ?causes haemolysis · Discontinue infusion 2-hrly & reassess
Vecuronium cost: cheap <£2/hr	0.08-0.1 mg/kg	50-80 μg/kg/hr (Intermittent injection preferred)	· EHL short · Prolonged in renal failure	· Stable CVS · Low potential for histamine release		· Non-depolarizing relaxant causing muscle paralysis · Anaphylaxis & prolonged effects noted · Neuromuscular monitoring recommended. · Boluses possibly better than infusion
Atracurium cost: medium <£5/hr	0.6 mg/kg	0.1-1 mg/kg/hr	· EHL short	· Stable CVS · Histamine released		· Non-depolarizing relaxant causing muscle paralysis · Anaphylaxis reported · Rapidly metabolized independent of liver/kidney · Used in cirrhosis & myasthenia gravis · A metabolite - laudanosine may accumulate causing seizures after some days

Bibliography

Aitkenhead AR. Analgesia and sedation in intensive care. Br J Anaesth 1989; 63: 96–106.

Bodenham A, Brownlie G, Dixon JS, and Park GR. Reversal of sedation by prolonged infusion of flumazenil. Anaesthesia 1988; 43: 367–368.

Carter JA, Arnold R, Yate PM, and Flynn PJ. Assessment of the Datex Relaxograph during anaesthesia and atracurium induced neuromuscular blockade. Br J Anaesth 1986; 58: 1447–1452.

Cohen AT and Kelly DR. Assessment of alfentanil by intravenous infusion as a long term sedation in intensive care. Anaesthesia 1987; 42: 545–548.

Committee on Safety of Medicines. Propofol—convulsion, anaphylaxis and delayed recovery from anaesthesia. 1989; 26.

Dobb GJ and Murphy DF. Sedation and analgesia during intensive care. In: Clin Anaesthesiol WB Saunders, 1985 Vol 3 (4): 1055.

Duthrie DJR and Nimmo WS. Adverse effects of opioid analgesic drugs. Br J Anaesth 1987; 59: 61–77.

Editorial. Sedation in the Intensive Care Unit. Lancet 1984; i: 1388–1389.

Kong KL, Willatts SM, and Prys-Roberts C. Isoflurane compared with midazolam for sedation in the Intensive Care Unit. Br Med J 1989; 298: 1277–1279.

Newman LH, McDonald JC, Wallace PG, and Ledingham I McA. Propofol infusion for sedation in intensive care. Anaesthesia 1987; 42: 929–937.

O'Sullivan GF and Park GR. The assessment of sedation in critically ill patients. Clin Intensive Care 1990; 1: 116–122.

Ramsay MAE, Savage TM, Simpson BAJ, and Goodwin R. Controlled sedation with alphaxolone-alphadone. Br Med J 1974; 2: 656–659.

Scott B, Beamish D, Hudson IN, and Jostell KG. Prolonged infusion of chlormethiazole in intensive care. Br J Anaesth 1980; 52: 541–545.

Sebel PS and Lowdon JD. Propofol: a new intravenous anaesthetic. Anesthesiology 1989; 71: 260–277.

Segredo V, Matthay MA, Sharma ML, Gruenke LD, Caldwell JE, and Millar RD. Prolonged neuromuscular blockade after long term administration of vecuronium in two critically ill patients. Anesthesiology 1990; 72: 566–70.

Shelly MP, Mendel L, and Park GR. Failure of critically ill patients to metabolise midazolam. Anaesthesia 1987; 42: 619–626.

Smith CL, Hunter JM, and Jones RS. Vecuronium infusions in patients with renal failure in an ITU. Anaesthesia 1987; 42: 387–393.

Yate PM, Thomas D, Short SM, Sebel PS, and Morton J. Comparison of infusion of alfentanil or pethidine for sedation of ventilated patients on the ITU. Br J Anaesth 1986; 58: 1091–1099.

1.5/1.6 Enteral nutrition/parenteral nutrition

Nutritional requirements
1. Resting energy expenditure can be calculated from VO_2 and VCO_2 using indirect calorimetry.
2. Energy requirements of critically ill patients vary widely. Standard formulae and tables both over- and under-estimate need.
3. Energy and nitrogen requirements are closely linked. A guide for the average male:

Metabolic rate	Energy requirement	Nitrogen requirement
Normal, resting	1500–2000 Cal/day	8–12 g/day
Increased	2000–2500 Cal/day	12–16 g/day
Severely increased	2000–3000 Cal/day	16–20 g/day

4. Highly catabolic patients retain less protein at the same administration rates compared to healthy individuals.
5. Vitamins and trace elements are essential although the exact requirements for critically ill patients are usually unknown.
6. Urinary excretion of urea over 24 hr can be used to estimate urinary nitrogen excretion.
   e.g.   [Nitrogren] (g) = 0.028 × [Urea] (mmol/24hr) + 2

Nutrient sources
1. A balanced amino acid solution should contain at least 40% essential amino acids and be rich in branched chain amino acids (BCAA).
2. BCAA infusions are expensive and shown to be beneficial only in burns and liver failure.
3. A balanced amino acid solution gives adequate protein synthesis if adequate non-protein energy is infused.
4. Patients with poor respiratory reserve or who are weaning from mechanical ventilation may benefit from a higher ratio of fat to carbohydrate.
5. A combination of glucose and fat emulsion for energy avoids the complications associated with infusion of either alone.
6. Gluconeogenesis is not effectively suppressed by exogenous glucose infusions in the critically ill.
7. Fat emulsions prevent essential fatty acid deficiency. At least 1000 ml 10% lipid emulsion is required weekly.
8. Administration of non-protein caloric substrates enhances the retention of exogenous amino acids.

Enteral versus parenteral feeding
1. Enteral nutrition is the best support of gastrointestinal mucosal integrity.
2. Improved bowel adaptation after resection.
3. Less risk of infection.
4. Absence of bowel sounds is not necessarily a contraindication to enteral feeding.

Complications of feeding
1. Diarrhoea associated with enteral feeding may be due to:
   - concomitant antibiotic therapy
   - inadequate bowel adaptation after prolonged disuse.
2. Avoiding catheter related sepsis in parenteral nutrition requires:

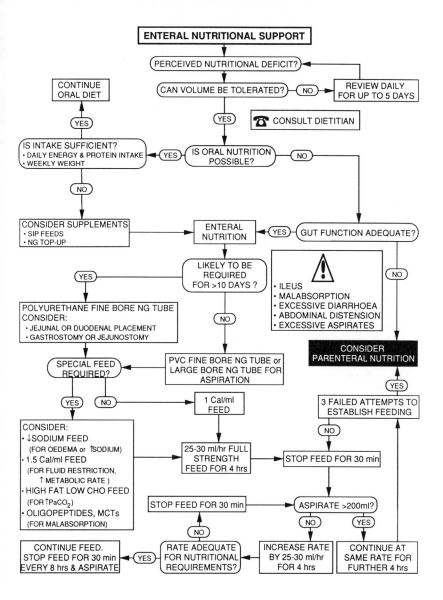

ENTERAL NUTRITIONAL SUPPORT

PERCEIVED NUTRITIONAL DEFICIT?

CAN VOLUME BE TOLERATED? → NO → REVIEW DAILY FOR UP TO 5 DAYS

CONTINUE ORAL DIET

YES

☎ CONSULT DIETITIAN

IS INTAKE SUFFICIENT?
• DAILY ENERGY & PROTEIN INTAKE
• WEEKLY WEIGHT
← YES — IS ORAL NUTRITION POSSIBLE? — NO →

NO

CONSIDER SUPPLEMENTS
• SIP FEEDS
• NG TOP-UP

ENTERAL NUTRITION ← YES — GUT FUNCTION ADEQUATE?

LIKELY TO BE REQUIRED FOR >10 DAYS?

⚠
• ILEUS
• MALABSORPTION
• EXCESSIVE DIARRHOEA
• ABDOMINAL DISTENSION
• EXCESSIVE ASPIRATES

NO

YES

NO

POLYURETHANE FINE BORE NG TUBE
CONSIDER:
• JEJUNAL OR DUODENAL PLACEMENT
• GASTROSTOMY OR JEJUNOSTOMY

CONSIDER PARENTERAL NUTRITION

YES

SPECIAL FEED REQUIRED? ← PVC FINE BORE NG TUBE or LARGE BORE NG TUBE FOR ASPIRATION

3 FAILED ATTEMPTS TO ESTABLISH FEEDING

NO

YES   NO

1 Cal/ml FEED

CONSIDER:
• ↓SODIUM FEED
(FOR OEDEMA or ↑SODIUM)
• 1.5 Cal/ml FEED
(FOR FLUID RESTRICTION, ↑ METABOLIC RATE)
• HIGH FAT LOW CHO FEED
(FOR ↑PaCO₂)
• OLIGOPEPTIDES, MCTs
(FOR MALABSORPTION)

25-30 ml/hr FULL STRENGTH FEED FOR 4 hrs → STOP FEED FOR 30 min

STOP FEED FOR 30 min → ASPIRATE >200ml?

NO   YES

CONTINUE FEED. STOP FEED FOR 30 min EVERY 8 hrs & ASPIRATE ← YES — RATE ADEQUATE FOR NUTRITIONAL REQUIREMENTS? ← INCREASE RATE BY 25-30 ml/hr FOR 4 hrs   CONTINUE AT SAME RATE FOR FURTHER 4 hrs

NO

- • strict asepsis when attending to catheter
- • dedicated central venous catheter
- • subcutaneous tunnel
- • experienced operator to insert catheter
- • experienced nurse to dress catheter
- • parenteral nutrition solutions compounded in pharmacy sterile unit.
3. Aspiration of enteral feed is a serious complication. Duodenal feeding may provide a safer route.
4. Glucose metabolism is impaired in the critically ill patient.
5. Non-oxidative metabolism of excess glucose leads to fatty infiltration of the liver.
6. Fine-bore NG tubes can be passed into the trachea inadvertently in both conscious and unconscious patients. Do chest X-ray after insertion.

Inadequate gastric emptying

1. The production of gastric juice (up to 125 ml/hr with a nasogastric tube in situ) should be taken into account when calculating the significance of nasogastric aspirates.
2. Large volume (> 200 ml) aspirates should not automatically lead to abolition of attempts to establish enteral feeding.
3. Large volume aspirates once enteral feeding has been established should be managed by:
   - Temporary reduction in feed volume.
   - Aspiration 4 hourly.
   - Further reductions in feed volume if large volume aspirates persist. The feed should only be stopped if large volume aspirates persist after 3 attempts to re-establish enteral feeding.
4. If nasogastric feeding must be halted due to large volume aspirates nasoduodenal or nasojejunal feeding should be considered early.
5. Opiates delay gastric emptying.
6. Metoclopramide may facilitate gastric emptying.

Bibliography

Al-Saady NM, Blackmore CM, and Bennett ED. High fat, low carbohydrate enteral feeding lowers $PaCO_2$ and reduces the period of ventilation in artificially ventilated patients. Intensive Care Med 1989; 15: 290–295.

American Medical Association. Guidelines for multivitamin preparations for parenteral use. JPEN 1979; 3: 258–262.

Askanazi J, Weissman C, Rosenbaum SH, Hyman AI, Milic-Emili J, and Kinney JM. Nutrition and the respiratory system. Crit Care Med 1982; 10: 163–172.

Burke JF, Wolfe RR, Mullany CJ, Mathews DE, and Bier DM. Glucose requirements following burn injury. Ann Surg 1979; 190: 274–285.

Elwyn DH. Nutritional requirements of adult surgical patients. Crit Care Med 1980; 8(1): 9–20.

Feldman EJ, Dowling RH, and McNaughton J, et al.. Effects of oral versus intravenous nutrition on intestinal adaptation after small bowel resection in the dog. Gastroenterology 1976; 70: 712–719.

MacFie J, Smith RC, and Hill GL. Glucose or fat as a non-protein energy source? A controlled clinical trial in gastroenterological patients requiring intravenous nutrition. Gastroenterology 1981; 80: 103–107.

Mann S, Westenskow DR, and Houtchens BA. Measured and predicted caloric expenditure in the critically ill. Crit Care Med 1985; 13: 173–177.

Schlichtig R and Ayres SM. Nutritional support of the critically ill. Year Book Medical Publishers, Chicago, 1988.

Vitamin and essential trace element recommendations during intravenous nutrition: theory and practice. Proc Nutr Soc 1986; 45: 383–390.

Weissmann C, Kemper M, Askanazi J, Hyman AI, and Kinney JM. Resting metabolic rate of the critically ill patient: measured versus predicted. Anesthesiology 1986; 64: 673–679.

Woolfson AMJ. Intravenous feeding—a review of current practice. Clin Nutr 1981; 4: 187–194.

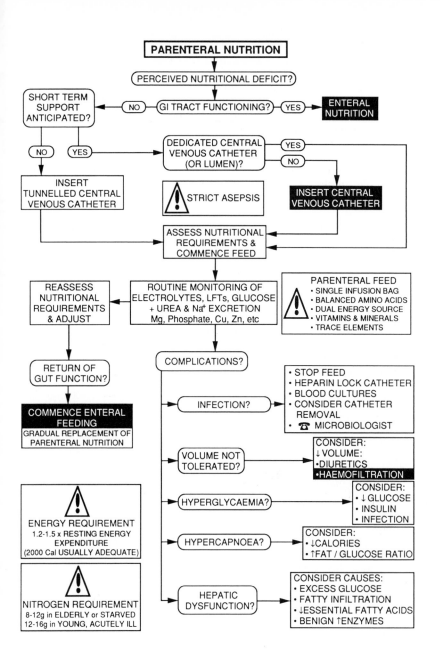

PARENTERAL NUTRITION

PERCEIVED NUTRITIONAL DEFICIT?

SHORT TERM SUPPORT ANTICIPATED? ← NO — GI TRACT FUNCTIONING? — YES → **ENTERAL NUTRITION**

NO / YES

DEDICATED CENTRAL VENOUS CATHETER (OR LUMEN)? — YES / NO

INSERT TUNNELLED CENTRAL VENOUS CATHETER

⚠ STRICT ASEPSIS

INSERT CENTRAL VENOUS CATHETER

ASSESS NUTRITIONAL REQUIREMENTS & COMMENCE FEED

REASSESS NUTRITIONAL REQUIREMENTS & ADJUST

ROUTINE MONITORING OF ELECTROLYTES, LFTs, GLUCOSE + UREA & Na⁺ EXCRETION Mg, Phosphate, Cu, Zn, etc

PARENTERAL FEED
⚠
• SINGLE INFUSION BAG
• BALANCED AMINO ACIDS
• DUAL ENERGY SOURCE
• VITAMINS & MINERALS
• TRACE ELEMENTS

RETURN OF GUT FUNCTION?

COMPLICATIONS?

COMMENCE ENTERAL FEEDING GRADUAL REPLACEMENT OF PARENTERAL NUTRITION

INFECTION?
• STOP FEED
• HEPARIN LOCK CATHETER
• BLOOD CULTURES
• CONSIDER CATHETER REMOVAL
• ☎ MICROBIOLOGIST

VOLUME NOT TOLERATED?
CONSIDER:
↓ VOLUME:
•DIURETICS
•HAEMOFILTRATION

⚠
ENERGY REQUIREMENT
1.2-1.5 x RESTING ENERGY EXPENDITURE
(2000 Cal USUALLY ADEQUATE)

HYPERGLYCAEMIA?
CONSIDER:
• ↓ GLUCOSE
• INSULIN
• INFECTION

HYPERCAPNOEA?
CONSIDER:
• ↓CALORIES
• ↑FAT / GLUCOSE RATIO

⚠
NITROGEN REQUIREMENT
8-12g in ELDERLY or STARVED
12-16g in YOUNG, ACUTELY ILL

HEPATIC DYSFUNCTION?
CONSIDER CAUSES:
• EXCESS GLUCOSE
• FATTY INFILTRATION
• ↓ESSENTIAL FATTY ACIDS
• BENIGN ↑ENZYMES

1.7: Pyrexia

Infection
1. Pus should be actively sought and drained.
2. Nosocomial (hospital-acquired) infections are frequently 'line-related' and/or Gram-negative and/or caused by resistant organisms.
3. Cultures should be taken before commencement of therapy.
4. Antibiotics may be responsible for unexplained pyrexias. If possible, stop therapy and reculture after 24 hrs.
5. Immunocompromised patients should receive antibiotic therapy if febrile.
6. Consider specific infections, e.g. malaria, TB, Legionnaire's.

Line changes
1. Venous and arterial lines should be changed at least every 5–6 days. Send line tips for culture.
2. Strict aseptic technique should be used for line insertions.

Hyperthermia
1. Hyperthermia is defined as a core temperature $> 40.5°C$.
2. Delirium occurs between $40–42°C$, coma above $42–43°C$.
3. The nervous system is most commonly and severely affected.
4. Signs:
   - confusion, delirium, convulsions, pupillary abnormalities, coma
   - tachycardia, ECG changes, initial high cardiac output, high VO_2 and low SVR, later cardiac failure
   - tachypnoea, respiratory alkalosis
   - loss of water, Na^+, K^+, Ca^{++}, Mg^{++} from excessive sweating
   - ATN, rhabdomyolysis
   - thrombocytopenia, DIC, haemolysis.

Rapid cooling
1. Rapid cooling should be instituted for core temperatures $>40.5°C$.
2. Remove clothing.
3. Use ice packs. Consider ice bath and bladder, gastric and peritoneal lavage with iced saline.
4. Wet skin, warm and circulate surrounding air to increase evaporative losses.
5. Consider phenothiazines (paralysis if ventilated) to stop shivering.
6. For malignant hyperpyrexia (q.v.) give dantrolene 1 mg/kg/min, repeated every 5 min to total 10 mg/kg. Consider using dantrolene for other causes. Avoid verapamil with dantrolene.
7. Minimize handling.
8. Stop active cooling when core temp $<39°C$.

Specific syndromes
1. Malignant hyperpyrexia:
   - related to use of volatile anaesthetic agents (NB halothane) and depolarizing muscle relaxants (suxamethonium). Also antidepressants and alcohol.
   - stop drug, cool actively, give dantrolene (as above), IPPV, high FiO_2, treat hyperkalaemia.
2. Neuroleptic malignant syndrome:
   - related to phenothiazines, butyrophenones and other antipsychotics.

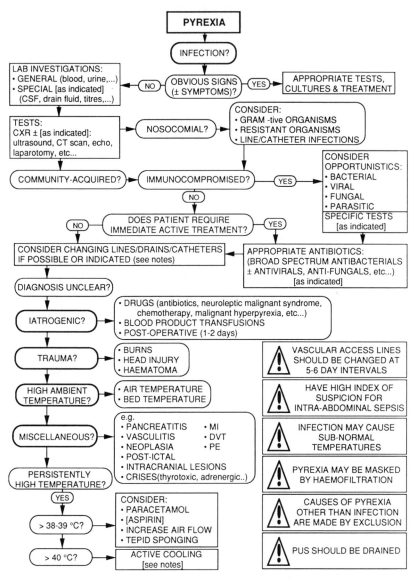

- hyperthermia, muscle rigidity, akinesia, impaired consciousness, autonomic dysfunction.
- recovery takes 1–2 weeks.
- treatment is based on stopping the offending agent, supportive care (active cooling, IPPV, fluid repletion, etc.) and specific agents, e.g. L−dopa+carbidopa, bromocriptine, dantrolene.

Bibliography

Aun C. Thermal syndromes. In: Intensive Care Manual 3rd Edition. Oh TE. (Ed.) Butterworths Sydney 1990 pp 470–3.

Gill RT, Kruse JA, Thill-Baharozian MC, and Carlson RW. Triple vs single lumen central venous catheters. A prospective study in a critically ill population. Arch Intern Med 1989; 149: 1139–1143.

2. Respiratory

2.1: Intermittent positive pressure ventilation

Guidelines for initial ventilator settings (adult)
- Minute volume   85–100 ml/kg/min
- Tidal volume   7–10 ml/kg
- Respiratory rate   12–18 bpm
- I:E ratio   1:1.5–1:2
- Airway pressure   < 40 cmH$_2$0

Dangers
1. At airway pressures >40 cmH$_2$O the incidence of pulmonary barotrauma increases.
2. In cases of severe bronchospasm or obstructive airways disease the respiratory rate may have to be reduced, the I:E ratio prolonged, and/or hypercapnia tolerated. It is vital to reassess the patient constantly NB increases in intrinsic PEEP (air-trapping).
3. For severe respiratory hypoxaemia reverse I:E ratios (in extreme cases up to 3:1 or 4:1), with a low flow rate, may be useful for oxygenation and avoidance of high airway pressures. The respiratory rate often needs to be reduced to avoid excessive increases in end-expiratory lung volume. Cardiac output may fall with this manoeuvre.

Modes
1. There are a variety of modes, some permitting respiratory efforts by the patient. Allowing the patient to breathe may:
   - reduce sedation requirements
   - re-train inspiratory muscles
   - reduce mean airway pressures.

   However, it is important to avoid condemning the patient to fruitless increases in respiratory work by using poorly designed circuitry. In this respect the use of continuous positive airways pressure (CPAP) and inspiratory pressure support appear promising.
2. For spontaneously breathing but intubated patients, T-piece circuits without CPAP should not be used for more than 30 minutes because of the loss of laryngeal PEEP and reduction in functional residual capacity.

Humidification
1. Endotracheal intubation bypasses the normal heat and moisture exchanging areas of the body. Artificial humidification is therefore necessary but imperfect.
2. Heated water systems are popular but can cause:
   - condensation with damage to mechanical parts of the circuitry
   - increased risk of nosocomial infection if not changed regularly
   - overheating due to electrical malfunction.
3. Heat and mositure exchangers (HME) are convenient to use but additional humidification is necessary. They are associated with an increased incidence of tube blockage. The Pall Ultipore HME also acts as a barrier to the passage of bacteria and is associated with a significant reduction in circuit contamination.
4. Both heated water systems and HMEs increase the resistance of the circuit.

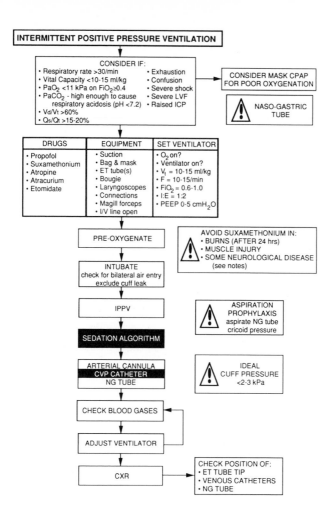

INTERMITTENT POSITIVE PRESSURE VENTILATION

CONSIDER IF:
- Respiratory rate >30/min
- Vital Capacity <10-15 ml/kg
- PaO_2 <11 kPa on FiO_2 ≥0.4
- $PaCO_2$ - high enough to cause respiratory acidosis (pH <7.2)
- Vd/Vt >60%
- Qs/Qt >15-20%
- Exhaustion
- Confusion
- Severe shock
- Severe LVF
- Raised ICP

CONSIDER MASK CPAP FOR POOR OXYGENATION

⚠ NASO-GASTRIC TUBE

DRUGS	EQUIPMENT	SET VENTILATOR
• Propofol	• Suction	• O_2 on?
• Suxamethonium	• Bag & mask	• Ventilator on?
• Atropine	• ET tube(s)	• V_t = 10-15 ml/kg
• Atracurium	• Bougie	• F = 10-15/min
• Etomidate	• Laryngoscopes	• FiO_2 = 0.6-1.0
	• Connections	• I:E = 1:2
	• Magill forceps	• PEEP 0-5 cmH_2O
	• I/V line open	

PRE-OXYGENATE

AVOID SUXAMETHONIUM IN:
⚠ • BURNS (AFTER 24 hrs)
• MUSCLE INJURY
• SOME NEUROLOGICAL DISEASE
(see notes)

INTUBATE
check for bilateral air entry
exclude cuff leak

IPPV

⚠ ASPIRATION PROPHYLAXIS
aspirate NG tube
cricoid pressure

SEDATION ALGORITHM

ARTERIAL CANNULA
CVP CATHETER
NG TUBE

⚠ IDEAL CUFF PRESSURE <2-3 kPa

CHECK BLOOD GASES

ADJUST VENTILATOR

CXR

CHECK POSITION OF:
• ET TUBE TIP
• VENOUS CATHETERS
• NG TUBE

Bibliography

Bersten AD, Rutten AJ, Vedig AE, and Skowronski A. Additional work of breathing imposed by endotracheal tubes, breathing circuits and intensive care ventilators. Crit Care Med 1989; 17: 671–677.

Branthwaite MA. Acute on chronic respiratory failure. Clin Anaesth 1985, 3: 831–847.

Brochard L, Pluskwa F, and Lemaire F. Improved efficacy of spontaneous breathing with inspiratory pressure support. Am Rev Respir Dis 1987; 136: 411–415.

Cullen DJ and Caldera DL. The incidence of ventilator-induced pulmonary barotrauma in critically ill patients. Anesthesiology 1979; 50: 185–190.

Kirby RR. Synchronised intermittent mandatory ventilation versus assist control: Just the facts, ma'am. Crit Care Med 1989; 17: 706–707.

Lin Es, Oh TE. Ventilation. Which mode? In: Balliere's Clinical Anaesthesiology Vol. 4 No. 2 Dobb GJ (Ed.) Balliere Tindall, London Philadelphia 1990; 471–473.

Martin C, Perrin G, Gevandain MJ, Saux P, and Gouin F. Heat and moisture exchangers and vaporising humidifiers in the Intensive Care Unit. Chest 1990; 97: 144–149.

Oh TE. Ventilation—matching man, mode and machine. Br J Hosp Med 1988; 40: 216–220.

Shelly MP, Lloyd GM, and Park GR. A review of the mechanism and method of humidification of inspired gases. Intensive Care Med 1988; 14: 1–9.

Stock MC, Downs JB, and Frolicher DA. Airway pressure release ventilation. Crit Care Med 1987; 15: 462–465.

2.2: Positive end-expiratory pressure (PEEP)

Benefits
1. PEEP increases functional residual capacity and lung compliance. It does not prevent or attenuate ARDS, reduce capillary leak or lung water, or hasten lung recovery.
2. Improvements in oxygenation are achieved by redistribution of lung fluid (from alveoli to interstitial spaces) and reversal of microatelectasis.
3. PEEP may increase cardiac output, particularly in LV failure states (see p. 36).
4. The level of 'optimal PEEP' is controversial. We believe that it is the lowest level that achieves an arterial oxygen saturation >90% allowing, wherever possible, reductions in FiO_2 (ideally to, or below, 0.6) though not at the expense of peak airway pressures >40 cmH_2O or significant falls in DO_2.

Risks
1. PEEP may increase peak and mean airway pressures and thus the risk of barotrauma. It may also increase intracranial pressure and hepatic congestion.
2. PEEP-related falls in cardiac output (q.v.) may outweigh any improvement in arterial oxygen saturation causing a fall in tissue oxygen delivery. Significant falls in DO_2 may occur at levels of PEEP <10 cmH_2O. More intensive haemodynamic monitoring is warranted if increasing PEEP > 5 cmH_2O.

Haemodynamics
1. PEEP usually reduces both left and right ventricular preload and increases RV afterload. Though PEEP may increase cardiac output in left ventricular failure, in most other cases it may cause a reduction in cardiac output. Furthermore, it may compromise a poorly functioning right ventricle.
2. Fluids and inotropes may restore cardiac output following an increase in PEEP. This should generally be avoided unless the increase in PEEP is necessary to maintain adequate arterial oxygenation. Caution should be exercised in patients with myocardial ischaemia.
3. Paradoxically, reductions in venous admixture with improvement in oxygenation may arise solely from reductions in cardiac output.
4. PEEP may shift the area of lung in which the pulmonary artery catheter tip is positioned from zone III to non-zone III. This is suggested by a rise in wedge pressure of at least half the increase in PEEP.

Intrinsic PEEP (PEEPi, auto-PEEP)
1. Intrinsic PEEP occurs in the presence of increased airflow resistance (e.g, emphysema, asthma) when the expiratory time is insufficient for complete exhalation. This leads to air trapping, CO_2 retention and increased airway pressures. It can be measured by temporarily occluding the expiratory port of the ventilator at end-expiration.
2. High levels of PEEPi can slow weaning by increasing the work of breathing. Use of extrinsic PEEP may overcome this.
3. Using high or reversed I:E ratios will often increase the level of intrinsic PEEP.

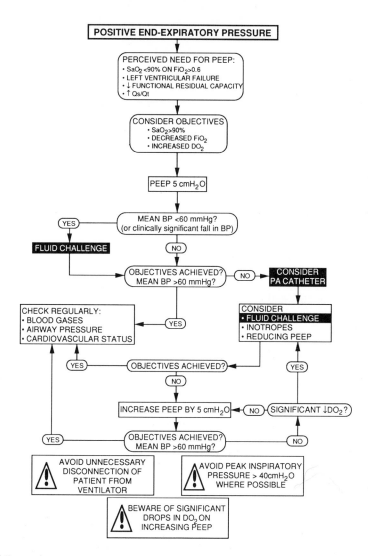

POSITIVE END-EXPIRATORY PRESSURE

PERCEIVED NEED FOR PEEP:
- SaO_2 <90% ON FiO_2 >0.6
- LEFT VENTRICULAR FAILURE
- ↓ FUNCTIONAL RESIDUAL CAPACITY
- ↑ Qs/Qt

CONSIDER OBJECTIVES
- SaO_2 >90%
- DECREASED FiO_2
- INCREASED DO_2

PEEP 5 cmH_2O

MEAN BP <60 mmHg?
(or clinically significant fall in BP) — YES → FLUID CHALLENGE

NO

OBJECTIVES ACHIEVED?
MEAN BP >60 mmHg? — NO → CONSIDER PA CATHETER

YES

CHECK REGULARLY:
- BLOOD GASES
- AIRWAY PRESSURE
- CARDIOVASCULAR STATUS

CONSIDER
- FLUID CHALLENGE
- INOTROPES
- REDUCING PEEP

OBJECTIVES ACHIEVED? — YES

NO

INCREASE PEEP BY 5 cmH_2O — NO → SIGNIFICANT ↓DO_2?

YES

OBJECTIVES ACHIEVED?
MEAN BP >60 mmHg? — YES / NO

⚠ AVOID UNNECESSARY DISCONNECTION OF PATIENT FROM VENTILATOR

⚠ AVOID PEAK INSPIRATORY PRESSURE > 40cmH_2O WHERE POSSIBLE

⚠ BEWARE OF SIGNIFICANT DROPS IN DO_2 ON INCREASING PEEP

Bibliography

Civetta JM. After quibbles and contrasts, concepts and caveats. Chest 1988; 93: 897–898.

Grace MP and Greenbaum DM. Cardiac performance in response to PEEP in patients with cardiac dysfunction. Crit Care Med 1982; 10: 358–360.

Schulman DS, Biondi JW, Matthay RA, Zaret BL, and Soufer R. Differing responses in right and left ventricular filling, loading and volumes during positive end-expiratory pressure. Am J Cardiol 1989; 64: 772–777.

Singer M and Bennett ED. Optimisation of positive end-expiratory pressure for maximal delivery of oxygen to tissues using oesophageal Doppler ultrasonography. Br Med J 1989; 298: 1350–1353.

Smith TC and Marini JJ. Impact of PEEP on lung mechanics and work of breathing in severe airflow obstruction. Am J Physiol 1988; 65: 1488–1499.

Tuxen DV. Detrimental effects of positive end-expiratory pressure during controlled mechanical ventilation of patients with severe airflow obstruction. Am Rev Respir Dis 1989; 140: 5–9.

Tyler DC. Positive end-expiratory pressure: a review. Crit Care Med 1983; 11: 300–308.

2.3: Weaning from intermittent positive pressure ventilation (IPPV)

1. Intermittent mandatory ventilation (IMV), continuous positive pressure ventilation (CPAP) and inspiratory pressure support (IPS) are all useful methods of weaning. However, there is no evidence that any of these techniques actually shorten the period of weaning.
2. Conventional weaning criteria are not infallible. In particular, respiratory mechanics criteria such as vital capacity and minute volume are significantly worse than oxygenation criteria such as PaO_2 and Qs/Qt in predicting successful weaning.
3. Weaning may be made more difficult by imposed restrictions to breathing, i.e. circuitry resistance, low gas flows and slow responses of inspiratory valves to patient demand. In particular, narrow-bore endotracheal tubes and connections may cause dramatic increases in inspiratory work. Inspiratory pressure support helps to overcome this.
4. Failure to wean is associated with:
   - an increased oxygen cost of breathing (as a percentage of total body oxygen uptake)
   - diaphragmatic (and intercostal) muscle fatigue
   - excessive superimposed respiratory work
   - poor nutrition
   - heart failure
   - low phosphate levels.
5. The major factor in successful weaning is the resolution of the precipitating illness. In addition, attention should be paid to the following:
   - improvement of nutritional state
   - eradication of sepsis
   - lowering of oxygen demands and CO_2 production
   - correction of fluid and electrolyte imbalance, magnesium and phosphate
   - treatment of heart failure and optimization of oxygen delivery
   - avoidance of high intra-abdominal pressures
   - provision of adequate analgesia.

Bibliography

Annat GJ, Viale JP, Deremey CP, Bouffard YM, Delafosse BX, and Motin JP. Oxygen cost of breathing and diaphragmatic pressure-time index. Measurement in patients with COPD during weaning with pressure support ventilation. Chest 1990; 98: 411–414.

Brochard L, Harf A, Lorino H, and Lemaire F. Inspiratory pressure support prevents diaphragmatic fatigue during weaning from mechanical ventilation. Am Rev Respir Dis 1989; 139: 513–521.

Brown DRG. Weaning patients from mechanical ventilation. Intensive Care Med 1984; 10: 55–59.

Dehaven CB, Hurst JM, and Branson RD. Evaluation of two different extubation criteria. Attributes contributing to success. Crit Care Med 1986; 14: 92–94.

Fiastro JF, Habib MP, Shon BY, and Campbell SC. Comparison of standard weaning parameters and the mechanical work of breathing in mechanically ventilated patients. Chest 1988; 94: 232–238.

Milic-Emili J. Is weaning an art or a science? Am Rev Respir Dis 1986; 134: 1107–1108.

Roussos C and Macklem PT. The respiratory muscles. N Engl J Med 1982; 307; 786–796.

Sahn SA, Lakshminarayan S, and Petty TL. Weaning from mechanical ventilation. JAMA 1976; 235: 2208–2212.

Tobin MJ. Predicting weaning outcome (Editorial). Chest 1988; 94: 227–228.

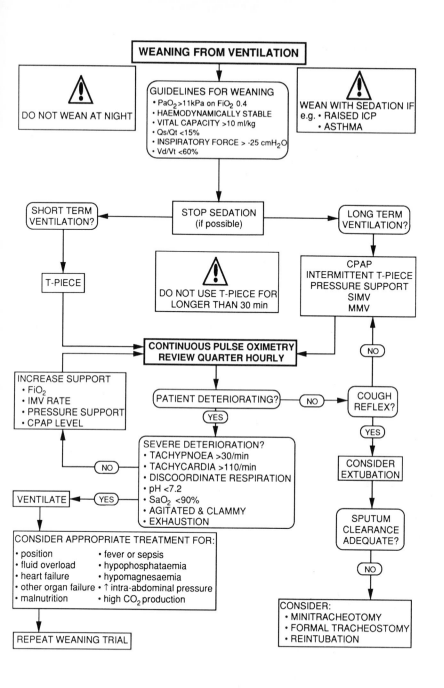

WEANING FROM VENTILATION

⚠ DO NOT WEAN AT NIGHT

GUIDELINES FOR WEANING
- $PaO_2 > 11$ kPa on FiO_2 0.4
- HAEMODYNAMICALLY STABLE
- VITAL CAPACITY >10 ml/kg
- Qs/Qt <15%
- INSPIRATORY FORCE > -25 cmH_2O
- Vd/Vt <60%

⚠ WEAN WITH SEDATION IF e.g. • RAISED ICP • ASTHMA

SHORT TERM VENTILATION? ← **STOP SEDATION** (if possible) → **LONG TERM VENTILATION?**

T-PIECE

⚠ DO NOT USE T-PIECE FOR LONGER THAN 30 min

CPAP INTERMITTENT T-PIECE PRESSURE SUPPORT SIMV MMV

CONTINUOUS PULSE OXIMETRY REVIEW QUARTER HOURLY

INCREASE SUPPORT
- FiO_2
- IMV RATE
- PRESSURE SUPPORT
- CPAP LEVEL

PATIENT DETERIORATING? —NO→ COUGH REFLEX?

YES

NO

SEVERE DETERIORATION?
- TACHYPNOEA >30/min
- TACHYCARDIA >110/min
- DISCOORDINATE RESPIRATION
- pH <7.2
- SaO_2 <90%
- AGITATED & CLAMMY
- EXHAUSTION

—NO→

VENTILATE ← YES

YES

CONSIDER EXTUBATION

CONSIDER APPROPRIATE TREATMENT FOR:
- position
- fluid overload
- heart failure
- other organ failure
- malnutrition
- fever or sepsis
- hypophosphataemia
- hypomagnesaemia
- ↑ intra-abdominal pressure
- high CO_2 production

SPUTUM CLEARANCE ADEQUATE?

NO

REPEAT WEANING TRIAL

CONSIDER:
- MINITRACHEOTOMY
- FORMAL TRACHEOSTOMY
- REINTUBATION

Viale JP, Annat G, Bertrand O, Godard J, and Motin J. Additional inspiratory work in intubated patients breathing with CPAP systems. Anesthesiology 1985; 63: 536–539.

2.4: Haemothorax/Pneumothorax

Management of pneumothorax
1. The differential diagnosis includes bullous emphysema, pneumatocoele and intra-pulmonary cyst. It is important to recognize these conditions and avoid a chest drain unless there is an assured pneumothorax. They are associated with pneumothoraces and require careful monitoring in the ventilated patient.
2. A pneumothorax may be missed on a supine chest X-ray since the lung edge or an absence of lung markings may not be seen. A hyperlucent lung field, loss of clarity of diaphragm outline, the 'deep sulcus' sign or a particularly clear cardiac margin are suggestive.
3. It may take the lung several weeks to re-expand with a conservatively treated pneumothorax.
4. The lung rarely re-expands on its own if a pneumothorax is associated with underlying diffuse lung disease.
5. All pneumothoraces should be drained in mechanically ventilated patients.
6. All pneumothoraces should be drained prior to any inter-hospital transfer.

Insertion of a chest drain
1. A chest drain of size 28 French or larger is required for a haemothorax whereas 20 French will suffice for a pure pneumothorax.
2. Insert via the Vth intercostal space in the mid axillary line.
3. Anaesthetize skin and pleura with 1% plain lignocaine. Ensure that air or fluid is aspirated.
4. Make a 1–1.5 cm skin crease incision and create a track with gloved finger (or artery forceps) to separate muscle fibres and to open pleura.
5. To avoid lung damage insert the drain with the trochar withdrawn. Angle and insert the drain to a correct position. Connect drain to water-filled bottle.
6. The chest drain should be directed towards the apex of the lung for a pneumothorax and towards the base of the lung for a haemothorax.
7. The drain should be well secured to the chest wall by properly placed sutures.
8. Always do a check chest X-ray to ensure correct placement and lung reinflation.

Management of the chest drain
1. Chest drains do not need to be clamped prior to removal or transport of the patient.
2. Chest drains may be removed in spontaneously breathing patients when the lung has re-expanded and there is no air leak.
3. A chest drain inserted for haemothorax may be removed during IPPV if there is no air leak and the lung has re-expanded.
4. The risk of lung entrapment and infarction is increased by the practice of chest drain milking and this should be avoided in pneumothorax.
5. Remove chest drain at end-expiration.

Complications of chest drainage
1. Morbidity associated with chest drainage may be up to 10%:
   - damage to intrathoracic viscera or intercostal vessels
   - septic complications
   - chest wall discomfort
   - impaired coughing
   - dislodgement of tube.

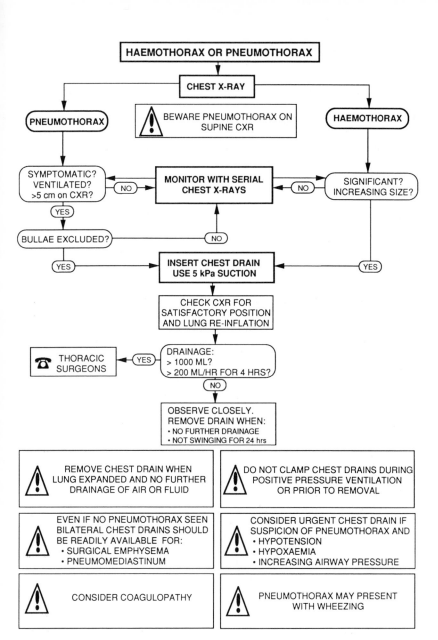

HAEMOTHORAX OR PNEUMOTHORAX

CHEST X-RAY

PNEUMOTHORAX

⚠ BEWARE PNEUMOTHORAX ON SUPINE CXR

HAEMOTHORAX

SYMPTOMATIC? VENTILATED? >5 cm on CXR? —NO→ **MONITOR WITH SERIAL CHEST X-RAYS** ←NO— SIGNIFICANT? INCREASING SIZE?

YES

BULLAE EXCLUDED? —NO→

YES

INSERT CHEST DRAIN USE 5 kPa SUCTION ←YES

CHECK CXR FOR SATISFACTORY POSITION AND LUNG RE-INFLATION

☎ THORACIC SURGEONS ←YES— DRAINAGE: > 1000 ML? > 200 ML/HR FOR 4 HRS?

NO

OBSERVE CLOSELY. REMOVE DRAIN WHEN:
• NO FURTHER DRAINAGE
• NOT SWINGING FOR 24 hrs

⚠ REMOVE CHEST DRAIN WHEN LUNG EXPANDED AND NO FURTHER DRAINAGE OF AIR OR FLUID

⚠ DO NOT CLAMP CHEST DRAINS DURING POSITIVE PRESSURE VENTILATION OR PRIOR TO REMOVAL

⚠ EVEN IF NO PNEUMOTHORAX SEEN BILATERAL CHEST DRAINS SHOULD BE READILY AVAILABLE FOR:
• SURGICAL EMPHYSEMA
• PNEUMOMEDIASTINUM

⚠ CONSIDER URGENT CHEST DRAIN IF SUSPICION OF PNEUMOTHORAX AND
• HYPOTENSION
• HYPOXAEMIA
• INCREASING AIRWAY PRESSURE

⚠ CONSIDER COAGULOPATHY

⚠ PNEUMOTHORAX MAY PRESENT WITH WHEEZING

Bibliography

Cummin A. Pneumothorax in the supine patient. Br Med J 1987; 295: 591–592.

Duncan C and Erickson R. Pressures associated with chest tube stripping. Heart Lung 1982; 11: 166–171.

Gordon R. The deep sulcus sign. Radiology 1980; 136: 25–27.

Gustman P, Yerger L, and Wanner A. Immediate cardiovascular effects of tension pneumothorax. Am Rev Respir Dis 1983; 127: 171–174.

Tocino IM, Miller MH and Fairfax WR. Distribution of pneumothorax in the supine and semi-recumbent critically ill adult. AJR 1985; 144: 901–905.

Wallis J and Wells F. Chest injuries: diagnosis and management. Care of the Critically Ill 1987; 3: 187–192.

2.5: Acute asthma

Severity
1. Mortality from asthma has increased and is often avoidable by early and aggressive therapy.
2. Two groups of patients with severe acute asthma may be identified:
   - insidious deterioration leading to exhaustion.
   - acute catastrophic bronchospasm with early asphyxiation.
3. Patients with marked 'morning dipping' are at risk of sudden acute catastrophic attacks.
4. The following features on admission suggest a requirement for mechanical ventilation:
   - pulsus paradoxus >30 mmHg.
   - heart rate >110/min.
   - PEFR >80 l/min.
   - previous severe asthma requiring mechanical ventilation.

Therapy
1. A high $PaCO_2$ in acute asthma is not, in general, a contraindication to high FiO_2 therapy.
2. Treatment of both bronchospasm and inflammation is required.
3. Patients who develop respiratory muscle fatigue may avoid mechanical ventilation with the cautious use of CPAP to reduce the work of breathing.
4. Cautious CPAP in patients with a raised FRC may not raise FRC further.
5. Careful attention to hydration is required to avoid mucus plugging.
6. If β_2 agonists, anticholinergics and theophyllines fail to secure improvement, the use of ketamine or halothane as a bronchodilator may be of use in mechanically ventilated cases.

Mechanical ventilation
1. Patients are likely to require additional humidification for removal of inspissated secretions.
2. Excessive air-trapping should be avoided since it is associated with:
   - increased risk of pneumothorax
   - poor left and right ventricular function by distortion of the RV and septum.
3. Pulmonary distension is reduced by accepting moderate hypercapnia:
   - low frequency ventilation (6–10/min).
   - low tidal volume (6–10 ml/kg).
   - low inspiratory flow rates.
   - prolonged expiratory time. ˏ
   - avoid high peak airway pressures (>50 cmH_2O).
   - adequate sedation ± muscle relaxants to overcome respiratory drive.

Bibliography
Benatar S. Fatal asthma. N Engl J Med 1986; 314: 423–429.
Branthwaite MA. The intensive care of asthma. Brit J Hosp Med 1985; 34: 331–338.
Broseghini C, Brandolese R, Rossi R, *et al*. Respiratory mechanics during the first day of mechanical ventilation in patients with pulmonary edema and chronic airway obstruction. Am Rev Respir Dis 1988; 138: 355–361.
Burney PGJ. Asthma mortality in England and Wales: evidence for a further increase, 1979–84. Lancet 1986; ii: 323–326.
Darioli R and Perret C. Mechanical controlled hypoventilation in status asthmaticus. Am Rev Respir Dis 1984; 129: 385–387.

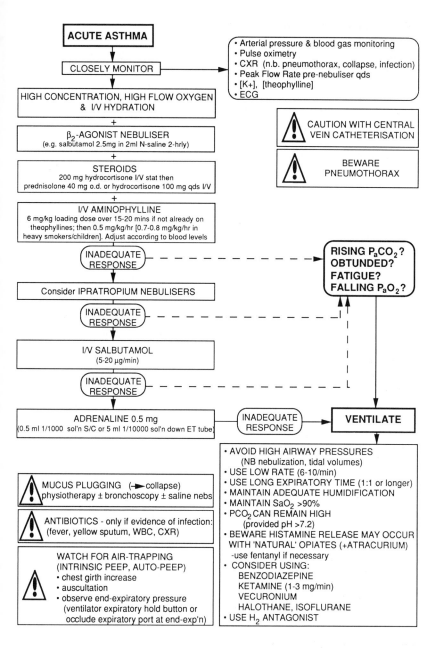

ACUTE ASTHMA

CLOSELY MONITOR →
- Arterial pressure & blood gas monitoring
- Pulse oximetry
- CXR (n.b. pneumothorax, collapse, infection)
- Peak Flow Rate pre-nebuliser qds
- [K+], [theophylline]
- ECG

HIGH CONCENTRATION, HIGH FLOW OXYGEN & I/V HYDRATION

+

β$_2$-AGONIST NEBULISER
(e.g. salbutamol 2.5mg in 2ml N-saline 2-hrly)

+

STEROIDS
200 mg hydrocortisone I/V stat then
prednisolone 40 mg o.d. or hydrocortisone 100 mg qds I/V

+

I/V AMINOPHYLLINE
6 mg/kg loading dose over 15-20 mins if not already on
theophyllines; then 0.5 mg/kg/hr [0.7-0.8 mg/kg/hr in
heavy smokers/children]. Adjust according to blood levels

INADEQUATE RESPONSE

Consider IPRATROPIUM NEBULISERS

INADEQUATE RESPONSE

I/V SALBUTAMOL
(5-20 µg/min)

INADEQUATE RESPONSE

ADRENALINE 0.5 mg
(0.5 ml 1/1000 sol'n S/C or 5 ml 1/10000 sol'n down ET tube)

INADEQUATE RESPONSE →

⚠ CAUTION WITH CENTRAL VEIN CATHETERISATION

⚠ BEWARE PNEUMOTHORAX

RISING P$_a$CO$_2$?
OBTUNDED?
FATIGUE?
FALLING P$_a$O$_2$?

VENTILATE

- AVOID HIGH AIRWAY PRESSURES
  (NB nebulization, tidal volumes)
- USE LOW RATE (6-10/min)
- USE LONG EXPIRATORY TIME (1:1 or longer)
- MAINTAIN ADEQUATE HUMIDIFICATION
- MAINTAIN SaO$_2$ >90%
- PCO$_2$ CAN REMAIN HIGH
  (provided pH >7.2)
- BEWARE HISTAMINE RELEASE MAY OCCUR
  WITH 'NATURAL' OPIATES (+ATRACURIUM)
  -use fentanyl if necessary
- CONSIDER USING:
  BENZODIAZEPINE
  KETAMINE (1-3 mg/min)
  VECURONIUM
  HALOTHANE, ISOFLURANE
- USE H$_2$ ANTAGONIST

⚠ MUCUS PLUGGING (→collapse)
physiotherapy ± bronchoscopy ± saline nebs

⚠ ANTIBIOTICS - only if evidence of infection:
(fever, yellow sputum, WBC, CXR)

⚠ WATCH FOR AIR-TRAPPING
(INTRINSIC PEEP, AUTO-PEEP)
- chest girth increase
- auscultation
- observe end-expiratory pressure
  (ventilator expiratory hold button or
  occlude expiratory port at end-exp'n)

Freedman AR and Lavietes MH. Energy requirements of the respiratory musculature in asthma. AM J Med 1986; 80: 215–222.

Hillman K. Pulmonary barotrauma. Clin Anaesthesiol 1985; 3: 877–898.

Jones D *et al.* Can we recognise very severe asthma? Thorax 1978; 33: 675.

Park GR, Manara AR, Mendel L, *et al.* Ketamine infusion. Its use as a sedative, inotrope and bronchodilator in a critically ill patient. Anaesthesia 1987; 42: 980–983.

Rea HH, Scragg R, Jackson R, Beaglehole R, Fenwick J, and Sutherland DC. A case-control study of deaths from asthma. Thorax 1986; 41: 833–839.

Shivaram U *et al.* Effects of continuous positive airway pressure in acute asthma. Respiration 1987; 52: 157–162.

2.6: Management of burn/smoke inhalation

Oxygen therapy in the presence of carbon monoxide or cyanide poisoning
1. If high O_2 concentrations are needed use an anaesthetic-type face mask and reservoir bag connected to a high flow O_2 supply. Ambu bags and disposable commercially available masks (other than the Venturi 60% mask) give low and unpredictable concentrations.
2. The half-life of carboxyhaemoglobin is 4 hours when breathing air and only 50 mins on 100% O_2.
3. Continue high concentration oxygen therapy until carboxyhaemoglobin levels <10%.

Poisoning
1. A high arterial PO_2 does not exclude carbon monoxide poisoning. Pulse oximetry and blood gas analyser calculations are misleading. Only *direct* measurement of oxyhaemoglobin (and carboxyhaemoglobin) by a haemoximeter will provide the correct answer.
2. Cyanide poisoning can only be diagnosed clinically in this situation as a blood cyanide level will take 3 hours to perform. Given the uncertainty of the diagnosis, the safest treatment is oxygen therapy and sodium thiosulphate (50 ml of 25% sodium thiosulphate I/V over 10 mins). In mild poisoning (no history of coma) oxygen alone will suffice.

Miscellaneous
1. Smoke inhalation causes thermal injury to larynx and pharynx only as hot gases have a low specific heat content. Steam has a much higher heat content and causes injury to the whole respiratory tract.
2. If there is any doubt about the patency of the airway early intubation is advised. Once a leak can be demonstrated (with the cuff deflated) the tube can be removed.
3. Fluid requirements may increase in burns and pulmonary injury.

Bibliography
Broome JR, Pearson RR, and Skrine H. Carbon monoxide poisoning, forgotten not gone! Br J Hosp Med 1988; 39: 298–305.
Clark CJ, Reid WH, Gilmore WH, and Campbell D. Mortality probability in victims of fire trauma: revised equation to include inhalation injury. Br Med J 1986; 292: 1303–1305.
Heimbach DM. Inhalation injuries. Ann Emerg Med 1988; 17: 1316–1320.
Herndon DN, Langner F, Thompson P, Linares HA, Stein M, and Traber DL. Pulmonary injuries in burned patients. Surg Clin N Am 1987; 67: 31–46.
Jones J, McMullen, MJ, and Dougherty J. Toxic smoke inhalation. Cyanide poisoning in fire victims. Am J Emerg Med 1987; 5: 317–321.
Langford R and Armstrong RF. Algorithm for managing injury from smoke inhalation. Br Med J 1989; 299: 902–904.
Navar PD, Saffle JR, and Warden GD. Effect of inhalation injury on fluid resuscitation requirements after thermal injury. Am J Surg 1985; 150: 716–720.
Venus B, Matsuda T, Copiozo JB, and Mathru M. Prophylactic intubation and continuous positive airway pressure in the management of inhalation injury in burns victims. Crit Care Med 1981; 9: 519–523.

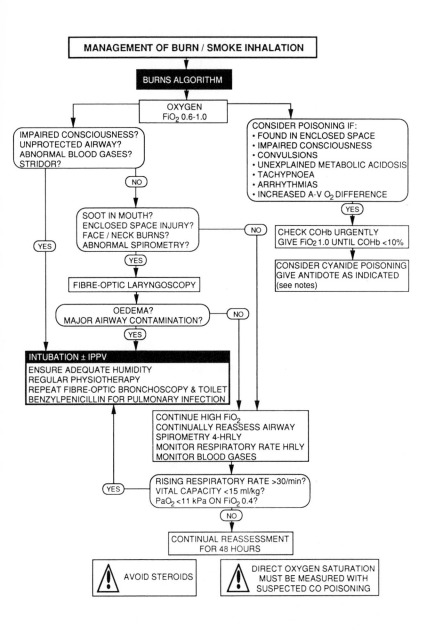

MANAGEMENT OF BURN / SMOKE INHALATION

BURNS ALGORITHM

OXYGEN
FiO₂ 0.6-1.0

IMPAIRED CONSCIOUSNESS?
UNPROTECTED AIRWAY?
ABNORMAL BLOOD GASES?
STRIDOR?

NO

CONSIDER POISONING IF:
• FOUND IN ENCLOSED SPACE
• IMPAIRED CONSCIOUSNESS
• CONVULSIONS
• UNEXPLAINED METABOLIC ACIDOSIS
• TACHYPNOEA
• ARRHYTHMIAS
• INCREASED A-V O₂ DIFFERENCE

YES

SOOT IN MOUTH?
ENCLOSED SPACE INJURY?
FACE / NECK BURNS?
ABNORMAL SPIROMETRY?

NO

YES

CHECK COHb URGENTLY
GIVE FiO₂ 1.0 UNTIL COHb <10%

CONSIDER CYANIDE POISONING
GIVE ANTIDOTE AS INDICATED
(see notes)

FIBRE-OPTIC LARYNGOSCOPY

OEDEMA?
MAJOR AIRWAY CONTAMINATION?

NO

YES

INTUBATION ± IPPV
ENSURE ADEQUATE HUMIDITY
REGULAR PHYSIOTHERAPY
REPEAT FIBRE-OPTIC BRONCHOSCOPY & TOILET
BENZYLPENICILLIN FOR PULMONARY INFECTION

CONTINUE HIGH FiO₂
CONTINUALLY REASSESS AIRWAY
SPIROMETRY 4-HRLY
MONITOR RESPIRATORY RATE HRLY
MONITOR BLOOD GASES

YES

RISING RESPIRATORY RATE >30/min?
VITAL CAPACITY <15 ml/kg?
PaO₂ <11 kPa ON FiO₂ 0.4?

NO

CONTINUAL REASSESSMENT
FOR 48 HOURS

⚠ AVOID STEROIDS

⚠ DIRECT OXYGEN SATURATION
MUST BE MEASURED WITH
SUSPECTED CO POISONING

3. Cardiovascular

3.1: Cardiac arrest

Cardiopulmonary resuscitation
1. The effect of cardiac massage is explained by two theories. Cardiac massage at 80–100/min is a compromise satisfying the cardiac compression and thoracic pump theories. Higher cardiac outputs are generated with higher rates if cardiac compression predominates and with longer compressions if thoracic pumping predominates.
2. There is a delay between defibrillation and return of the ECG trace. CPR should continue during this period.
3. End tidal CO_2 can be used to monitor the adequacy of cardiopulmonary resuscitation.
4. Fixed dilated pupils cannot be used as a prognostic sign during CPR.

Adrenaline
1. The α (vasoconstrictor) effects of adrenaline predominate during resuscitation thus maintaining aortic diastolic blood pressure and coronary and cerebral perfusion.
2. At least 10 $\mu g/kg/5$ min is required for optimal effect in all cardiac arrests, irrespective of rhythm.

Bicarbonate
1. Sodium bicarbonate may exacerbate intracellular and respiratory acidosis.
2. Whereas effective CPR may correct the cause of the metabolic acidosis, sodium bicarbonate does not.
3. If an adequate circulation is difficult to establish, sodium bicarbonate may temporarily correct a potentially lethal pH.

Calcium
1. Calcium prevents reperfusion of ischaemic brain.
2. Cytoplasmic calcium accumulation is associated with cell death.

Bibliography
Chamberlain DA. Advanced life support. Br Med J 1989; 299: 446–450.
Cingolani HE, Mattiazzi AR, Blesa ES, and Gonzalez NC. Contractility in isolated mammalian heart muscle after acid base changes. Circ Res 1970; 26: 269–278.
Dembo DH. Calcium in advanced life support. Crit Care Med 1981; 9: 358–359.
Graf H, Leach W, and Arieff AI. Evidence for a detrimental effect of bicarbonate therapy in hypoxic lactic acidosis. Science 1985; 227: 754–756.
Halperin HR, Tsitlk JE, Guerci AD, et al. Determinants of blood flow to vital organs during cardiopulmonary resuscitation in dogs. Circulation 1986; 73: 539–550.
Katz AM and Reuter H. Cellular calcium and cardiac cell death. Am J Cardiol 1979; 44: 188–190.
Maier GW, Tyson GS, Olsen CO, et al. The physiology of external cardiac massage: High impulse cardiopulmonary resuscitation. Circulation 1984; 70: 86–101.
Marsden AK. Basic life support. Br Med J 1989; 299: 442–445.
Michael JR, Guerci AD, Koehler RC, et al. Mechanisms by which epinephrine augments cerebral and myocardial perfusion during cardiopulmonary resuscitation in dogs. Circulation 1984; 69: 822–835.
Narins RG and Cohen JJ. Bicarbonate therapy for organic acidosis: The case for its continued use. Ann Intern Med 1987; 106: 615–618.
Renekov L. Calcium antagonist drugs—myocardial preservation and reduced vulnerability to ventricular fibrillation during CPR. Crit Care Med 1981; 9: 360–361.
Schleien CL, Berkowitz ID, Traystman R, and Rogers MC. Controversial issues in cardiopulmonary resuscitation. Anesthesiology 1989; 71: 133–149.

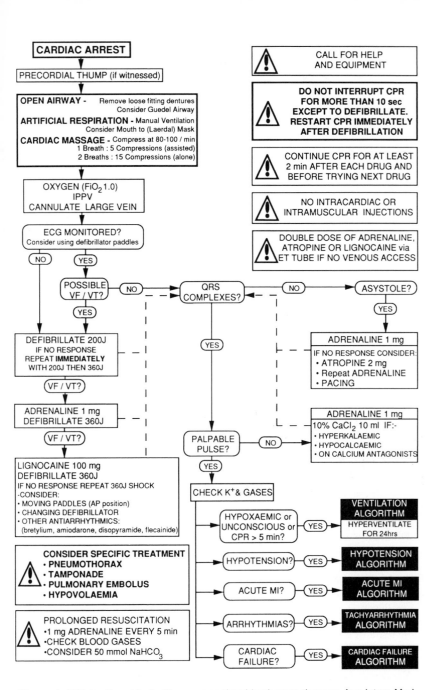

Stacpoole PW. Lactic acidosis: The case against bicarbonate therapy. Ann Intern Med 1986; 105: 276–279.

Standards for cardiopulmonary resuscitation (CPR) and emergency cardiac care (ECC). JAMA 1986; 255: 2905–2932.

Weil MH, Rackow EC, Trevino R, *et al*. Difference in acid-base state between venous and arterial blood during cardiopulmonary resuscitation. N Engl J Med 1986; 315: 153–156.

Yakaitis RW, Otto CW, and Blitt CD. Relative importance of α and β adrenergic receptors during resuscitation. Crit Care Med 1979; 7: 293–296.

3.2: Hypotension

Circulatory management
1. The circulating volume must be corrected before using catecholamines for hypotension.
2. A normal blood pressure does not guarantee an adequate cardiac output.
3. Catecholamines should be titrated against predefined physiological end-points.
4. Support of the circulation aims to maintain tissue oxygenation:
   - increases in cardiac output increase oxygen transport.
   - increases in blood pressure facilitate perfusion of the microvasculature.
5. Critically ill patients require higher than normal levels of oxygen transport to maintain a higher metabolic rate.
6. Increasingly complex regimens for treatment of shock generally require an increased complexity of monitoring.

Effects of catecholamines
1. Critically ill patients, with high circulating levels of endogenous catecholamines, may have down-regulated receptors and may thus require very high doses.
2. The predominant effects of catecholamines are determined by their receptor activity:

Catecholamine	Predominant receptor(s)	Predominant effects
Dobutamine	β_1	↑↑ Stroke Vol, ↑ Heart rate
Dopamine (low dose)	Dopaminergic	↑ Splanchnic and renal perfusion
Adrenaline	β_1, α_1 and β_1	↑ Heart rate, ↑ Stroke Vol, Vasoconstriction
Noradrenaline	α_1	Vasoconstriction
Isoprenaline	β_1 and β_2	↑↑ Heart rate, vasodilatation

3. Dobutamine and isoprenaline may reduce pulmonary vascular resistance.
4. Dobutamine may cause vasodilatation due to β_2 effects.
5. Dobutamine causes less tachycardia than isoprenaline.
6. Tachycardia increases myocardial oxygen demand severely.

The use of vasopressors in hypotension
1. Vasopressors may reduce cardiac output via the increase in peripheral resistance.
2. SVR may apparently increase due to a reduction in CO rather than an increase in BP.
3. Vasopressors should be titrated against the blood pressure rather than the SVR.
4. Vasopressors should be used in the smallest dose which achieves an acceptable blood pressure.
5. Although 60 mmHg is a minimum mean pressure to aim for, hypotension is best treated relative to pre-illness values.

Bibliography
Edwards JD, Brown CS, Nightingale P, Slater R, and Farragher EB. Use of survivors' cardio-respiratory values as therapeutic goals in septic shock. Crit Care Med 1989; 17; 1098–1103.

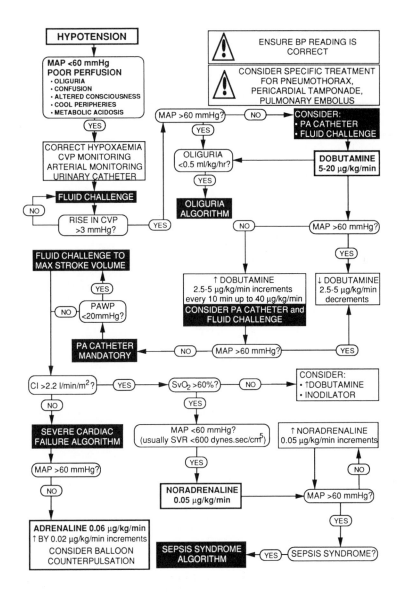

Maekawa K, Liang CS, and Hood WB. Comparison of dobutamine and dopamine in acute myocardial infarction. Effects of systemic hemodynamics, plasma catecholamines, blood flows and infarct size. Circulation 1983; 67: 750–759.

Meadows D, Edwards JD, Wilkins RG, and Nightingale P. Reversal of intractable septic shock with norepinephrine therapy. Crit Care Med 1988; 16: 663–666.

Shoemaker WC, Appel PL, Kram HB, Waxman K, and Lee TS. Prospective trial of supranormal values of survivors as therapeutic goals in high-risk surgical patients. Chest 1988; 94: 1176–1186.

Tuttle DR, Pollock GD, Todd G, MacDonald B, Tust R, and Dusenberry W. The effect of doubutamine on cardiac oxygen balance, regional blood flow and infarction severity after coronary artery narrowing in dogs. Circ Res 1977; 41: 357–364.

Vincent JL, Reuse C, and Kahn RJ. Effects on right ventricular function of a change from dopamine to dobutamine in critically ill patients. Crit Care Med 1988; 16: 659–662.

3.3: Tachyarrhythmia

1. Treat if there is (impending) haemodynamic compromise.
2. Treat the cause of arrhythmias whenever possible.
3. Drugs used for chronic treatment of arrhythmias may not be appropriate acutely.
4. Many patients requiring antiarrhythmic therapy acutely will not require long term therapy.
5. The risks of digitalis toxicity are higher in critically ill patients with poor excretory function.
6. Pacing should be available prior to treatment of digoxin induced tachyarrhythmias.
7. Unsuccessful digitalis therapy increases the risks of side effects from other antiarrhythmic therapy.

Management of atrial and supraventricular tachyarrhythmias
1. Verapamil usually works immediately.
2. Hypotension with verapamil usually responds to intravenous calcium.
3. Hypotension with amiodarone is often secondary to the solvent and is avoided by slow intravenous bolus injection.
4. The use of verapamil is contraindicated if the patient is β blocked.

Management of ventricular tachyarrhythmias
1. The use of verapamil in misdiagnosed ventricular tachycardia may cause life threatening hypotension.
2. Intravenous amiodarone can successfully revert otherwise intractable ventricular tachycardia.
3. Amiodarone can be used in combination with other therapy for ventricular tachycardia.
4. Overpacing is indicated if a pacing wire is *in situ* or if drug therapy fails.

Bibliography
Faniel R and Schoenfeld Ph. Efficacy of iv amiodarone in converting rapid atrial fibrillation and atrial flutter to sinus rhythm in intensive care patients. Eur Heart J 1983; 4: 180–185.
Iberti TJ, Benjamin E, Paluch TA, Gentili DR, and Gabrielson GV. Use of constant infusion verapamil for the treatment of postoperative supraventricular tachycardia. Crit Care Med 1986; 14: 283–284.
Installe E, Clabecq M. Gadisseux P, and Charles S. Evaluation of intravenous amiodarone in the treatment of various arrhythmias following heart surgery involving extracorporeal circulation. J Thorac Cardiovasc Surg 1981; 81: 301–308.
Munoz A, Karila P, Gallay P, Zettelmeier F, Messner P, Mery M, and Grolleau R. A randomized hemodynamic comparison of intravenous amiodarone with and without Tween 80. Eur Heart J 1988; 9: 142–148.
Reiter MJ, Shand DG, Aanonsen LM, *et al*. Pharmacokinetics of verapamil: Experience with a sustained intravenous infusion regimen. Am J Cardiol 1982; 50: 716–721.
Saksena S and Kesselbrenner MB. Current perspectives in management of ventricular tachyarrhythmias with intravenous and oral amiodarone. Clin Prog Electrophysiol Pacing 1986; 4: 382–394.
Sung RJ, Shapiro WA, Shen EN, Morady F, and Davis J. Effects of verapamil on ventricular tachycardias possibly caused by reentry, automaticity and triggered activity. J Clin Invest. 1983; 72: 350–360.

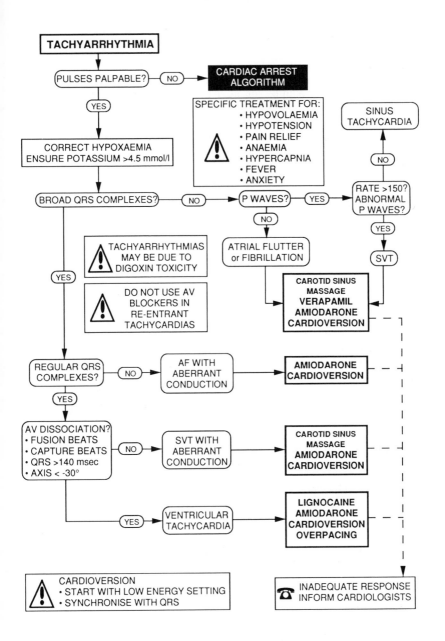

TACHYARRHYTHMIA

PULSES PALPABLE? — NO → **CARDIAC ARREST ALGORITHM**

YES

CORRECT HYPOXAEMIA
ENSURE POTASSIUM >4.5 mmol/l

SPECIFIC TREATMENT FOR:
• HYPOVOLAEMIA
• HYPOTENSION
• PAIN RELIEF
• ANAEMIA
• HYPERCAPNIA
• FEVER
• ANXIETY

SINUS TACHYCARDIA

BROAD QRS COMPLEXES? — NO → P WAVES? — YES → RATE >150? ABNORMAL P WAVES?

NO

YES

TACHYARRHYTHMIAS MAY BE DUE TO DIGOXIN TOXICITY

DO NOT USE AV BLOCKERS IN RE-ENTRANT TACHYCARDIAS

ATRIAL FLUTTER or FIBRILLATION

SVT

CAROTID SINUS MASSAGE VERAPAMIL AMIODARONE CARDIOVERSION

REGULAR QRS COMPLEXES? — NO → AF WITH ABERRANT CONDUCTION → **AMIODARONE CARDIOVERSION**

YES

AV DISSOCIATION?
• FUSION BEATS
• CAPTURE BEATS
• QRS >140 msec
• AXIS < -30°

NO → SVT WITH ABERRANT CONDUCTION → **CAROTID SINUS MASSAGE AMIODARONE CARDIOVERSION**

YES → VENTRICULAR TACHYCARDIA → **LIGNOCAINE AMIODARONE CARDIOVERSION OVERPACING**

CARDIOVERSION
• START WITH LOW ENERGY SETTING
• SYNCHRONISE WITH QRS

☎ INADEQUATE RESPONSE INFORM CARDIOLOGISTS

3.4: Severe cardiac failure

Respiratory management
1. High concentration, high flow oxygen therapy is mandatory.
2. Early ventilation should be considered in severe cardiac failure:
   - Reduced work of breathing.
   - Reduced metabolic rate.
   - Diaphragmatic failure associated with inadequate cardiac output.
   - Reduced cardiac distension due to reduced venous return.
   - Allows the use of sedatives and anxiolytics.
   - Reduced left ventricular afterload.
3. PEEP may improve gas exchange in severe pulmonary oedema but should not be allowed to reduce cardiac output.

Fluid management
1. Caution should be exercised in the use of frusemide.
2. Patients rarely die of pulmonary oedema in heart failure; rather they die of organ hypoperfusion.
3. Fluid overload is not usually a feature of acute heart failure; fluid is maldistributed so excessive diuresis should be avoided.
4. Cardiac performance in severe heart failure is worsened by compensatory vasoconstriction. Diuretics will cause further vasoconstriction by hypovolaemia.
5. Any beneficial effects of frusemide in acute heart failure without fluid overload are due to early and transient venodilatation.
6. The use of closely monitored (PA catheter) colloidal infusions may increase cardiac performance with an additional reduction in peripheral resistance.

Peripheral circulatory management
1. Nitrates produce an improved cardiac performance in severe heart failure by:
   - decreased peripheral resistance
   - reduced cardiac distension by reduced venous return
   - possibly increased coronary perfusion.
2. Hypotension is not a problem with nitrates if cardiac filling pressures are maintained.
3. Normalizing a raised SVR is an important method of improving cardiac efficiency.

Catecholamines
1. Catecholamines should be used sparingly in acute heart failure since they increase myocardial oxygen demand and decrease myocardial cAMP.
2. Inotropic catecholamines should not be titrated against blood pressure; rather they should be titrated against measures of adequacy of the circulation:
   - SvO_2 >60%
   - Normal blood lactate
   - Cardiac index >2.2 l/min/m^2
3. Vasopressor catecholamines are usually not required in acute heart failure since peripheral resistance is usually high.
4. Phosphodiesterase inhibitors (e.g. enoximone) may restore catecholamine sensitivity by reducing cAMP depletion.

Monitoring
1. PAWP of 15–18 mmHg is usually adequate but may need to be higher in chronic cardiac failure.
2. A fall in SvO_2 is the earliest indicator of circulatory inadequacy.

Bibliography

Aubier M, Trippenbach T, and Roussos C. Respiratory muscle fatigue during cardiogenic shock. J Appl Physiol 1981; 51: 499–508.

Bussman W and Schupp D. Effect of sublingual nitroglycerin in emergency treatment of severe pulmonary edema. Am J Cardiol 1978; 41: 931–936.

Creamer J, Edwards JD, and Nightingale P. Haemodynamic and oxygen transport variables in cardiogenic shock following acute myocardial infarction and their response to treatment. Am J Cardiol 1990; 65: 1297–1306.

Dikshit K, Vyden JK, Forrester JS, Chatterjee K, Prakash R, and Swan HJC. Renal and extrarenal hemodynamic effects of furosemide in congestive heart failure after acute myocardial infarction. N Engl J Med 1973; 288: 1087–1090.

Forrester JS, Diamond G. McHugh T, and Swan HJC. Filling pressures in the right and left sides of the heart in acute myocardial infarction. A reappraisal of central venous pressure monitoring. N Engl J Med 1971; 285: 190–192.

Nelson GIC, Ahuja RC, Silke B, Hussain M, and Taylor SH. Haemodynamic advantages of isosorbide dinitrate over frusemide in acute heart failure following myocardial infarction. Lancet 1983; i: 730–732.

Nelson GIC, Silke B, Forsyth DR, Verma SP, Hussain M, and Taylor SH. Hemodynamic comparison of primary venous or arteriolar dilatation and the subsequent effect of furosemide in left ventricular failure after acute myocardial infarction. Am J Cardiol 1983; 52: 1036–1040.

Singer M and Bennett ED. Intravenous nitrates in severe left ventricular failure with hypotension. Crit Care Med 1989; 17: S134.

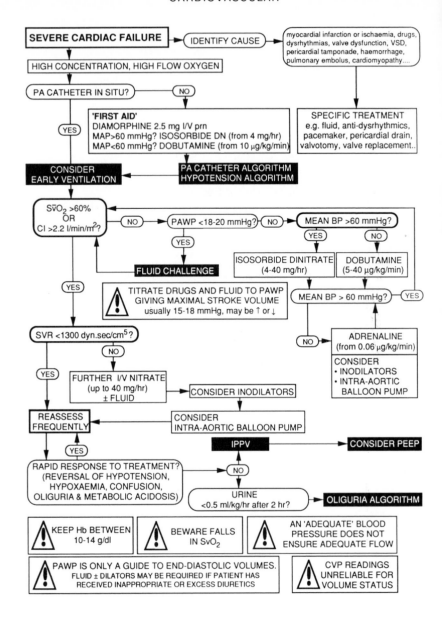

SEVERE CARDIAC FAILURE → IDENTIFY CAUSE → myocardial infarction or ischaemia, drugs, dysrhythmias, valve dysfunction, VSD, pericardial tamponade, haemorrhage, pulmonary embolus, cardiomyopathy....

HIGH CONCENTRATION, HIGH FLOW OXYGEN

PA CATHETER IN SITU? — NO

'FIRST AID'
DIAMORPHINE 2.5 mg I/V prn
MAP>60 mmHg? ISOSORBIDE DN (from 4 mg/hr)
MAP<60 mmHg? DOBUTAMINE (from 10 µg/kg/min)

SPECIFIC TREATMENT
e.g. fluid, anti-dysrhythmics, pacemaker, pericardial drain, valvotomy, valve replacement..

YES → CONSIDER EARLY VENTILATION ← PA CATHETER ALGORITHM / HYPOTENSION ALGORITHM

$S\bar{v}O_2$ >60% OR CI >2.2 l/min/m²? — NO → PAWP <18-20 mmHg? — NO → MEAN BP >60 mmHg?

YES → FLUID CHALLENGE

YES → ISOSORBIDE DINITRATE (4-40 mg/hr)

NO → DOBUTAMINE (5-40 µg/kg/min)

⚠ TITRATE DRUGS AND FLUID TO PAWP GIVING MAXIMAL STROKE VOLUME usually 15-18 mmHg, may be ↑ or ↓

MEAN BP > 60 mmHg? — YES

SVR <1300 dyn.sec/cm⁵? — NO

NO → ADRENALINE (from 0.06 µg/kg/min)

CONSIDER
• INODILATORS
• INTRA-AORTIC BALLOON PUMP

YES → FURTHER I/V NITRATE (up to 40 mg/hr) ± FLUID → CONSIDER INODILATORS

REASSESS FREQUENTLY ← CONSIDER INTRA-AORTIC BALLOON PUMP

IPPV → CONSIDER PEEP

YES → RAPID RESPONSE TO TREATMENT? (REVERSAL OF HYPOTENSION, HYPOXAEMIA, CONFUSION, OLIGURIA & METABOLIC ACIDOSIS) — NO → URINE <0.5 ml/kg/hr after 2 hr? → OLIGURIA ALGORITHM

⚠ KEEP Hb BETWEEN 10-14 g/dl

⚠ BEWARE FALLS IN SvO_2

⚠ AN 'ADEQUATE' BLOOD PRESSURE DOES NOT ENSURE ADEQUATE FLOW

⚠ PAWP IS ONLY A GUIDE TO END-DIASTOLIC VOLUMES. FLUID ± DILATORS MAY BE REQUIRED IF PATIENT HAS RECEIVED INAPPROPRIATE OR EXCESS DIURETICS

⚠ CVP READINGS UNRELIABLE FOR VOLUME STATUS

Notes

3.5: Myocardial infarction

Thrombolysis
1. Effective up to 24 hours. Preferably within 4–6 hours from onset of symptoms.
2. Revascularization arrhythmias are common post-thrombolysis. Of these over 90% are benign and do not require treatment.
3. Contraindications to thrombolysis:

Absolute contraindications	Relative contraindications
active G–I bleeding	traumatic CPR
aortic dissection	major surgery in last 10 days
neurosurgery/head injury/CVA within 2/12	past history of G–I bleeding
intracranial neoplasm/aneurysm	recent obstetric delivery
proliferative diabetic retinopathy	prior organ biopsy
	serious trauma
	severe hypertension
	bleeding diathesis
	(previous puncture of major vessel)

4. Streptokinase is the current first-line therapy unless the patient has had a previous reaction to it or has received it within the previous 6 months. If so use tissue plasminogen activator (rTPA).
5. Arterial and/or central venous cannulation should not be delayed if clinically indicated. It should be performed by an experienced operator, *avoiding* the subclavian route.
6. Thrombolysis may be given even when relative contraindications (e.g. hypotension, anterior MI) exist, when the mortality risk from MI outweighs the risk of bleeding.
7. Allergic reactions to thrombolytic therapy should be treated:
   - Stop streptokinase.
   - Hydrocortisone 100 mg I/V.
   - Chlorpheniramine 10 mg I/V.
   - Ranitidine 50 mg I/V.
   - Restart after 30 min.
8. If severe bleeding occurs during thrombolysis, or urgent surgery is required:
   - Stop infusion.
   - Give aprotonin 500,000 units over 10 min, then 200,000 units over 4hr.
   - OR give tranexamic acid 10 mg/kg repeated after 6–8 hr.
   - plus FFP (though must give with aprotonin or tranexamic acid).
9. Little information available on pregnancy and menstruation; rTPA generally preferable.
10. Hypotension:
    - prior to thrombolysis?—rTPA preferable.
    - during/after thrombolysis? Reduce/temporarily stop; support circulation and aim to continue.
11. There is no added benefit to aspirin by giving heparin.

Drug doses

1. Diamorphine	2.5 mg I/V. Repeat PRN + anti-emetic
2. Streptokinase	1.5 million units in 100ml 0.9% saline I/V over 1 hr
3. TPA	100mg I/V over 3 hr (10mg bolus, 50mg over 1 hr, 40 mg over 2 hr)
4. APSAC	30 units I/V over 5 min
5. Aspirin	150 mg. P/O od
6. Atenolol	50 mg P/O od (increase to 100mg od if not hypotensive & HR > 70 bpm)
7. Propranolol	10–40 mg P/O qds (titrate to HR of 60 bpm)
8. Isosorbide dinitrate	2–40 mg/hr I/V
9. GTN	10–200 μg/min I/V
10. Diltiazem	60 mg P/O tds
11. Nifedipine	5–10 mg S/L or P/O tds
12. Atropine	0.3 mg I/V. Repeated to maximum of 2 mg
13. Isoprenaline	1–4 μg/min
14. Lignocaine	1 mg/kg I/V bolus then 2–4 mg/min (1% solution contains 10mg/ml)
15. Amiodarone	5 mg/kg over 15 min then infused up to 15 mg/kg/day in 5% glucose via central vein. (In emergency: 150–300 mg in 10–20ml 5% glucose over 1–2 min)
16. Magnesium	$MgSO_4$ 20 mmol I/V over 2–3 hr

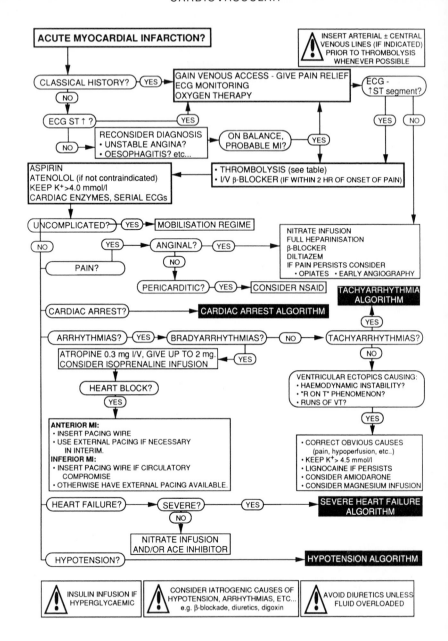

ACUTE MYOCARDIAL INFARCTION?

⚠ INSERT ARTERIAL ± CENTRAL VENOUS LINES (IF INDICATED) PRIOR TO THROMBOLYSIS WHENEVER POSSIBLE

CLASSICAL HISTORY? — YES → GAIN VENOUS ACCESS - GIVE PAIN RELIEF / ECG MONITORING / OXYGEN THERAPY → ECG - ↑ST segment?

NO

ECG ST ↑ ? — YES

NO → RECONSIDER DIAGNOSIS
• UNSTABLE ANGINA?
• OESOPHAGITIS? etc... → ON BALANCE, PROBABLE MI? — YES

YES / NO

ASPIRIN
ATENOLOL (if not contraindicated)
KEEP K⁺>4.0 mmol/l
CARDIAC ENZYMES, SERIAL ECGs

← • THROMBOLYSIS (see table)
• I/V β-BLOCKER (IF WITHIN 2 HR OF ONSET OF PAIN)

UNCOMPLICATED? — YES → MOBILISATION REGIME

NO

PAIN? — YES → ANGINAL? — YES → NITRATE INFUSION
FULL HEPARINISATION
β-BLOCKER
DILTIAZEM
IF PAIN PERSISTS CONSIDER
• OPIATES • EARLY ANGIOGRAPHY

NO

PERICARDITIC? — YES → CONSIDER NSAID

TACHYARRHYTHMIA ALGORITHM

CARDIAC ARREST? → **CARDIAC ARREST ALGORITHM**

YES

ARRHYTHMIAS? — YES → BRADYARRHYTHMIAS? — NO → TACHYARRHYTHMIAS?

ATROPINE 0.3 mg I/V, GIVE UP TO 2 mg.
CONSIDER ISOPRENALINE INFUSION — YES

NO

VENTRICULAR ECTOPICS CAUSING:
• HAEMODYNAMIC INSTABILITY?
• "R ON T" PHENOMENON?
• RUNS OF VT?

HEART BLOCK?

YES

YES

ANTERIOR MI:
• INSERT PACING WIRE
• USE EXTERNAL PACING IF NECESSARY IN INTERIM.
INFERIOR MI:
• INSERT PACING WIRE IF CIRCULATORY COMPROMISE
• OTHERWISE HAVE EXTERNAL PACING AVAILABLE.

• CORRECT OBVIOUS CAUSES
(pain, hypoperfusion, etc..)
• KEEP K⁺> 4.5 mmol/l
• LIGNOCAINE IF PERSISTS
• CONSIDER AMIODARONE
• CONSIDER MAGNESIUM INFUSION

HEART FAILURE? → SEVERE? — YES → **SEVERE HEART FAILURE ALGORITHM**

NO

NITRATE INFUSION
AND/OR ACE INHIBITOR

HYPOTENSION? → **HYPOTENSION ALGORITHM**

⚠ INSULIN INFUSION IF HYPERGLYCAEMIC

⚠ CONSIDER IATROGENIC CAUSES OF HYPOTENSION, ARRHYTHMIAS, ETC...
e.g. β-blockade, diuretics, digoxin

⚠ AVOID DIURETICS UNLESS FLUID OVERLOADED

Notes

Miscellaneous

1. Any suspicion of a low output state should result in active treatment.
2. An *early* cardiological opinion should be sought in patients with low output states as balloon angioplasty or early surgery may improve the prognosis in this high-risk group.
3. Diuretics are rarely needed if the patient does not have fluid overload or chronic heart failure. They may increase cardiac work by vasoconstriction secondary to volume depletion and activation of the renin-angiotensin-aldosterone system.
4. Hypotension may be iatrogenic, e.g. excessive diuretics, β-blockers.
5. Low dose nitrate infusions only cause hypotension in the underfilled patient.
6. Right ventricular infarction may result in underfilling of the left ventricle. Fluid loading and/or inotropic support is then indicated and a pulmonary artery catheter should be considered for precise management.
7. Patients developing post-infarction ventricular failure should ideally be commenced on ACE inhibitors; this does however await large study confirmation.
8. Unless contraindicated, atenolol 5 mg I/V should be given within two hours of onset of pain.
9. Caution should be exercised when giving non-steroidal anti-inflammatory drugs such as indomethacin as this may result in fluid retention.
10. When anginal-type pain persists post-infarction and does not settle on medical therapy, early angiography ± angioplasty ± early bypass grafting should be considered. Neither surgical technique will enhance resting ventricular function though it may prevent further deterioration.
11. Rarer causes of post-infarction heart failure and low output states should be considered, e.g. chordal rupture.
12. A modified Bruce exercise test should be performed prior to discharge.
13. The patient should ideally attend cardiac rehabilitation classes.

Bibliography

AIMS trial study group. Effect of intravenous APSAC on mortality after acute myocardial infarction: preliminary report of a placebo-controlled trial. Lancet 1988; i: 545–549.

Caplin JL. Acute right ventricular infarction. Br Med J 1989; 299: 459–460.

Chatterjee K, Swan HJC, Kaushik VS, Jobin G, Magnusson P, and Forrester JS. Effects of vasodilator therapy for severe pump failure in acute myocardial infarction on short-term and late prognosis. Circulation 1976; 53: 797–802.

Chesebro JH, Knatterud G, Roberts *et al.* Thrombolysis in myocardial infarction (TIMI) trial, Phase I: a Comparison between intravenous tissue plasminogen activator and intravenous streptokinase. Circulation 1987; 76: 142–154.

Gillespie TA, Ambos HD, Sobel BE, and Roberts R. Effects of dobutamine in patients with acute myocardial infarction. Am J Cardiol 1977; 39: 588–594.

Gruppo Italiano per lo studio della streptochinasi nell'infarco miocardico (GISSI). Effectiveness of intravenous thrombolytic treatment in acute myocardial infarction. Lancet 1986; i: 397–402.

Gruppo Italiano per lo studio della streptochinasi nell'infarco miocardico. GISSI–2. A factorial randomised trial of alteplase versus streptokinase and heparin versus no heparin among 12490 patients with acute myocardial infarction. Lancet 1990; 336: 65–71.

Johnson SA, Scanlon PJ, Loeb HS, Moran JM, Pifarre R, and Gunnar RM. Treatment of cardiogenic shock in myocardial infarction by intra-aortic balloon counterpulsation and surgery. Am J Med 1977; 62: 687–692.

Lee L, Bates ER, Pitt B, Walton JA, Laufer N, and O'Neill WW. Percutaneous transluminal coronary angioplasty improves survival in acute myocardial infarction complicated by cardiogenic shock. Circulation 1988; 78: 1345–1351.

Pfeffer MA, Lamas GA, Vaughan DE, Parisi AF, and Braunwald E. Effect of captopril on progressive ventricular dilatation after anterior myocardial infarction. N Engl J Med 1988; 319: 80–86.

Randomised trial of intravenous streptokinase, oral aspirin, both, or neither among 17187 cases of suspected acute myocardial infarction: ISIS–2. Lancet 1988; ii: 349–360.

Yusuf S, Peto R, Lewis J, Collins R, and Sleight P. Beta-blockade during and after myocardial infarction: an overview of the randomised trials. Prog Cardiovasc Dis 1985; 27: 335–371.

4. Renal

4.1: Oliguria

Urinary tract obstruction
1. Exclude a blocked catheter (common).
2. An ultrasound is suitable to exclude obstruction in the ICU patient.
3. An intravenous urogram is hazardous in renal failure.

Urinalysis
1. It is mandatory to test for protein, sugar and blood.
2. Granular casts indicate glomerular disease.
3. WBC (infection) and crystals (obstruction) may provide useful diagnostic information.

Chemistry
1. A low urinary sodium (<20 mmol/l) suggests hypovolaemia. It may be spuriously high after diuretics. It may be misleadingly low in burns, some phases of urinary tract obstruction, hepatorenal syndrome.
2. Differentiation of hypovolaemia from parenchymal renal failure:

	hypovolaemia (pre-renal)	renal
urine osmolality (mOsm/kg)	>500	<400
urinary Na	<20	>40
urine:plasma creatinine	>40	<20
fractional excretion Na	<1	>2

In practice, the important steps are to exclude obstruction, correct hypovolaemia and hypotension, and treat the cause. Chemical ratios are rarely needed.

Treatment
1. Correct hypovolaemia before giving drugs.
2. Diuretics may have a place in prophylaxis in some situations. Not shown to affect rate of recovery of established acute renal failure.
3. There is only limited evidence of the value of dopamine as prophylaxis against renal failure in man. There is no convincing evidence of its value in established renal failure.
4. Drug dosages should be modified in acute renal failure, NB:
   - antibiotics (aminoglycosides, penicillins. . .)
   - sedatives
   - muscle relaxants
   - digoxin.
5. Bigger doses of diuretics may be needed in long-term diuretic takers.

Miscellaneous
1. Consider rhabdomyolysis and myoglobinuria in oliguric patients, particularly if unconscious (e.g. from overdose) or involved in crush injury. Measure plasma CPK, urinary myoglobin. Treat by forced alkaline diuresis (with mannitol) and bicarbonate to keep urine pH $\geq$5.
2. If anuria persists, remove urinary catheter.

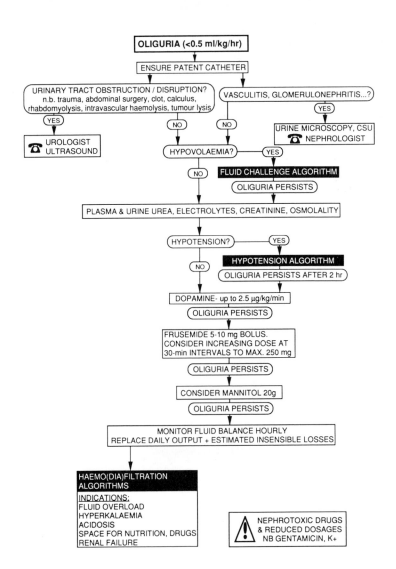

```
              ┌─────────────────────────┐
              │ OLIGURIA (<0.5 ml/kg/hr) │
              └─────────────────────────┘
                          ↓
              ┌─────────────────────────┐
              │ ENSURE PATENT CATHETER  │
              └─────────────────────────┘
```

OLIGURIA (<0.5 ml/kg/hr)

ENSURE PATENT CATHETER

URINARY TRACT OBSTRUCTION / DISRUPTION?
n.b. trauma, abdominal surgery, clot, calculus,
rhabdomyolysis, intravascular haemolysis, tumour lysis

VASCULITIS, GLOMERULONEPHRITIS...?

YES

YES

URINE MICROSCOPY, CSU
☎ NEPHROLOGIST

☎ UROLOGIST
ULTRASOUND

NO   NO

HYPOVOLAEMIA?   YES

NO

FLUID CHALLENGE ALGORITHM

OLIGURIA PERSISTS

PLASMA & URINE UREA, ELECTROLYTES, CREATININE, OSMOLALITY

HYPOTENSION?   YES

NO

HYPOTENSION ALGORITHM

OLIGURIA PERSISTS AFTER 2 hr

DOPAMINE- up to 2.5 µg/kg/min

OLIGURIA PERSISTS

FRUSEMIDE 5-10 mg BOLUS.
CONSIDER INCREASING DOSE AT
30-min INTERVALS TO MAX. 250 mg

OLIGURIA PERSISTS

CONSIDER MANNITOL 20g

OLIGURIA PERSISTS

MONITOR FLUID BALANCE HOURLY
REPLACE DAILY OUTPUT + ESTIMATED INSENSIBLE LOSSES

HAEMO(DIA)FILTRATION
ALGORITHMS
INDICATIONS:
FLUID OVERLOAD
HYPERKALAEMIA
ACIDOSIS
SPACE FOR NUTRITION, DRUGS
RENAL FAILURE

⚠ NEPHROTOXIC DRUGS
& REDUCED DOSAGES
NB GENTAMICIN, K+

Bibliography

Amerio A, Corabelli P, Campese VM, and Massry SG. Acute renal failure. Advances in experimental medicine and biology Vol. 212. Plenum Press, New York, London 1986.

Bihari D and Neild G. Acute renal failure in the intensive care unit. Springer-Verlag, Berlin 1990.

Gabow PA, Kaehry WD, and Kelleher SP. The spectrum of rhabdomyolysis. Medicine 1982; 61: 141–152.

Miller TR. Urinary diagnostic indices in acute renal failure: a prospective study. Ann Intern Med 1978; 89: 57–50.

Polson RJ, Park GR, Lindop MJ, Farman JV, Calne RY, and Williams R. The prevention of renal impairment in patients undergoing liver grafting by infusion of low dose dopamine. Anaesthesia 1987; 42: 15–19.

Prescribing in renal impairment. British National Formulary Vol. 19, Department of Health, London 1990; 19–27.

Schrier RW. Renal and electrolyte disorders. 3rd edition. Little, Brown, Boston, Toronto 1986.

4.2: Haemodiafiltration (anticoagulation)

1. Anticoagulation is often problematic—frequent checks of clotting useful.
2. Anticoagulation may not be necessary in patients with severe coagulopathy.
3. Consider prostacyclin if patient has ARDS.

Heparin-induced thrombocytopenia
1. Heparin flush of arterial lines may perpetuate the thrombocytopenia.
2. Switch to prostacyclin.
3. Consider low molecular wt heparins.

Bibliography
Ansell J, Slepchuk N Jr, Kumar R, Lopez A, Southard L, and Deykin D. Heparin induced thrombocytopenia: a prospective study. Thromb Haemostasis 1980; 43: 61–65.
Sinclair ME. Anticoagulation for extracorporeal systems. In: Update in Intensive Care & Emergency Medicine No. 5 Ed. Vincent J-L. Springer-Verlag Berlin 1988 pp. 693–695.
Weston MJ. Anticoagulation and extracorporeal circuits: the role of prostacyclin. In: Acute renal failure in the intensive care unit. Eds. Bihari D, Neild G. Springer-Verlag Berlin 1990; 337–347.

HAEMOFILTRATION

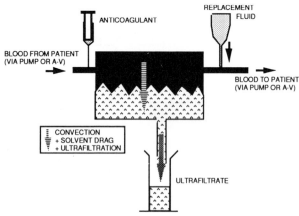

HAEMODIAFILTRATION

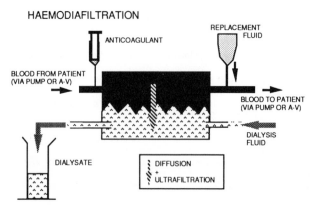

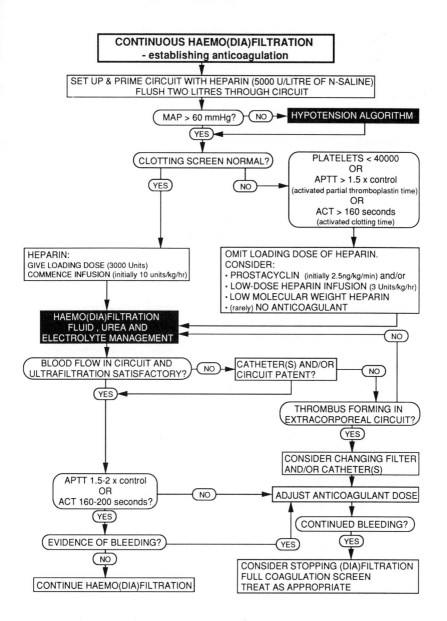

CONTINUOUS HAEMO(DIA)FILTRATION
- establishing anticoagulation

SET UP & PRIME CIRCUIT WITH HEPARIN (5000 U/LITRE OF N-SALINE)
FLUSH TWO LITRES THROUGH CIRCUIT

MAP > 60 mmHg? — NO → HYPOTENSION ALGORITHM

YES

CLOTTING SCREEN NORMAL?

YES   NO →

PLATELETS < 40000
OR
APTT > 1.5 x control
(activated partial thromboplastin time)
OR
ACT > 160 seconds
(activated clotting time)

HEPARIN:
GIVE LOADING DOSE (3000 Units)
COMMENCE INFUSION (initially 10 units/kg/hr)

OMIT LOADING DOSE OF HEPARIN.
CONSIDER:
• PROSTACYCLIN (initially 2.5ng/kg/min) and/or
• LOW-DOSE HEPARIN INFUSION (3 Units/kg/hr)
• LOW MOLECULAR WEIGHT HEPARIN
• (rarely) NO ANTICOAGULANT

HAEMO(DIA)FILTRATION
FLUID , UREA AND
ELECTROLYTE MANAGEMENT   ← NO

BLOOD FLOW IN CIRCUIT AND
ULTRAFILTRATION SATISFACTORY? — NO → CATHETER(S) AND/OR
CIRCUIT PATENT? — NO →

YES

THROMBUS FORMING IN
EXTRACORPOREAL CIRCUIT?

YES

CONSIDER CHANGING FILTER
AND/OR CATHETER(S)

APTT 1.5-2 x control
OR
ACT 160-200 seconds? — NO → ADJUST ANTICOAGULANT DOSE

YES

CONTINUED BLEEDING?

EVIDENCE OF BLEEDING? — YES →   YES

NO

CONSIDER STOPPING (DIA)FILTRATION
FULL COAGULATION SCREEN
TREAT AS APPROPRIATE

CONTINUE HAEMO(DIA)FILTRATION

4.3: Continuous haemodiafiltration (fluid, urea and electrolyte management)

General
1. Indications:
   ● fluid overload
   ● renal failure
   ● acidosis
   ● hyperkalaemia
   ● drug overdose
2. 'Hypercatabolic' is difficult to define. Consider when plasma creatinine is rising >100 μmol/l/day and urea >20 mmol/l/day. Haemodiafiltration indicated when hypercatabolic. When area <20 mmol/l and/or creatinine <200 μmol/l, consider change from haemodiafiltration to haemofiltration.
3. On haemodiafiltration, a dialysate flow of 1 litre/hr is usually sufficient. There is no value in exceeding 2 litre/hr.

Mechanical problems
1. Filters should last for a few days. Signs of filter failure include decreasing ultrafiltrate volumes and increasing creatinine and urea (if hypercatabolism ruled out).
2. If catheter is obstructed check and/or alter position. 'Kicking' of lines may be a sign of catheter obstruction. Consider exchanging lumens or, possibly, predilution.
3. Warm the replacement fluid to prevent hypothermia. Alternatively, cool for hyperthermia.

Metabolic
1. Check drug dosages and amend as necessary. Measure drug levels if possible (approx. creatinine clearance on haemofiltration 16 ml/min, on diafiltration 25 ml/min).
2. Acidosis may occur using lactate replacement fluid in liver failure and acetate containing fluid in hypoperfusion states (particularly muscle). Bicarbonate replacement fluid is available though unsuitable for some machines/filters.
3. Alkalosis may occur as excess buffer is present in the replacement fluid. Amino acid losses may occur through the filter.
4. Potassium can be added to the replacement bag to give a concentration of 4–5 mmol/l. Phosphate addition may be required. Check levels.

Bibliography
Burchardi H. Hemofiltration. In: Update in Intensive Care & Emergency Medicine No. 8. Vincent J-L (Ed.) Springer-Verlag, Berlin 1989; 340–347.
Lavar A, Saccaggi A, Ronco C, Belledonne M, Glabman S, and Bosch J. Continuous arteriovenous hemofiltration in the critically ill patient. Ann Intern Med 1989; 99: 455–460.
Miller R, Kingswood C, Bullen C, Cohen S. Renal replacement therapy in the ICU: the role of continuous arteriovenous haemodialysis. Br J Hosp Med 1990; 43: 354–362.
Schetz M, Lauwers PM, and Ferdinande P. Extracorporeal treatment of acute renal failure in the intensive care unit: a critical view. Intensive Care Med. 1989; 15: 349–357.
Stevens PE, Davies SP, Brown EA, Riley B, Gower PE, and Kox W. Continuous arteriovenous haemodialysis in critically ill patients. Lancet 1988; ii: 150–152.
Wendon J, Smithies M, Sheppard M, Bullen C, Tinker J, and Bihari D. Continuous high volume venous-venous haemofiltration in acute renal failure. Intensive Care Med 1989; 15: 358–363.

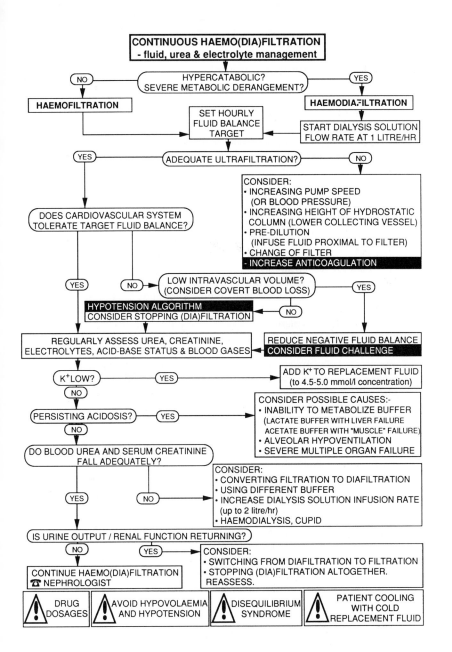

CONTINUOUS HAEMO(DIA)FILTRATION
- fluid, urea & electrolyte management

NO ⟵ HYPERCATABOLIC? SEVERE METABOLIC DERANGEMENT? ⟶ YES

HAEMOFILTRATION

SET HOURLY FLUID BALANCE TARGET

HAEMODIAFILTRATION

START DIALYSIS SOLUTION FLOW RATE AT 1 LITRE/HR

YES ⟵ ADEQUATE ULTRAFILTRATION? ⟶ NO

CONSIDER:
• INCREASING PUMP SPEED (OR BLOOD PRESSURE)
• INCREASING HEIGHT OF HYDROSTATIC COLUMN (LOWER COLLECTING VESSEL)
• PRE-DILUTION (INFUSE FLUID PROXIMAL TO FILTER)
• CHANGE OF FILTER
- INCREASE ANTICOAGULATION

DOES CARDIOVASCULAR SYSTEM TOLERATE TARGET FLUID BALANCE?

YES   NO

LOW INTRAVASCULAR VOLUME? (CONSIDER COVERT BLOOD LOSS) ⟶ YES

NO

HYPOTENSION ALGORITHM
CONSIDER STOPPING (DIA)FILTRATION

REGULARLY ASSESS UREA, CREATININE, ELECTROLYTES, ACID-BASE STATUS & BLOOD GASES

REDUCE NEGATIVE FLUID BALANCE
CONSIDER FLUID CHALLENGE

K^+LOW? ⟶ YES ⟶ ADD K^+ TO REPLACEMENT FLUID (to 4.5-5.0 mmol/l concentration)

NO

PERSISTING ACIDOSIS? ⟶ YES

CONSIDER POSSIBLE CAUSES:-
• INABILITY TO METABOLIZE BUFFER (LACTATE BUFFER WITH LIVER FAILURE ACETATE BUFFER WITH "MUSCLE" FAILURE)
• ALVEOLAR HYPOVENTILATION
• SEVERE MULTIPLE ORGAN FAILURE

NO

DO BLOOD UREA AND SERUM CREATININE FALL ADEQUATELY?

YES   NO

CONSIDER:
• CONVERTING FILTRATION TO DIAFILTRATION
• USING DIFFERENT BUFFER
• INCREASE DIALYSIS SOLUTION INFUSION RATE (up to 2 litre/hr)
• HAEMODIALYSIS, CUPID

IS URINE OUTPUT / RENAL FUNCTION RETURNING?

NO   YES

CONSIDER:
• SWITCHING FROM DIAFILTRATION TO FILTRATION
• STOPPING (DIA)FILTRATION ALTOGETHER.
REASSESS.

CONTINUE HAEMO(DIA)FILTRATION
☎ NEPHROLOGIST

⚠ DRUG DOSAGES

⚠ AVOID HYPOVOLAEMIA AND HYPOTENSION

⚠ DISEQUILIBRIUM SYNDROME

⚠ PATIENT COOLING WITH COLD REPLACEMENT FLUID

5. Metabolic

5.1/5.2: Hypo- and hypernatraemia

Hyponatraemia
1. Symptoms include nausea, vomiting, headache, muscular twitching, obtundation, psychosis, convulsions and coma.
2. In chronic hyponatraemia correction should not exceed 0.5 mmol/l/hr in the first 24 hr and 0.3 mmol/l/hr thereafter.
3. In acute hyponatraemia the ideal rate of correction is controversial. It is generally agreed that elevations in plasma sodium level can be faster, but no greater than 20 mmol/l/day. If doubt exists as to speed of onset of hyponatraemia, treat as if acute hyponatraemia.
4. A plasma sodium level of 125–130 mmol/l is a reasonable target to aim for in initial correction of both acute and chronic states. Attempts to rapidly achieve normo- or hypernatraemia should be avoided. Neurological complications e.g. central pontine myelinolysis, osmotic demyelination syndrome are related to the degree of correction and (in chronic hyponatraemia) the rate. Women are more prone to these complications.
5. Equations that calculate excess water are unreliable. It is safer to perform frequent estimations of plasma sodium levels.
6. Hypertonic saline infusions may be dangerous in the elderly or those with impaired cardiac function. An alternative is to use frusemide with replacement of urinary sodium (and potassium) losses each 2–3 hr. Thereafter simple water restriction is usually sufficient. NB many patients achieve normonatraemia by spontaneous water diuresis.
7. Use isotonic solutions for reconstituting drugs, parenteral nutrition, bladder irrigation, etc. . .
8. Severe hyponatraemia has been corrected safely by continuous arterio-venous haemofiltration.

Type	Aetiology	Urinary Na Conc'n
ECF VOLUME DEPLETION	RENAL LOSSES: diuretic excess, mineralocorticoid deficiency, salt-losing nephritis, renal tubular acidosis, osmotic diuresis (glucose, urea, mannitol)	>20 mmol/l
	EXTRA-RENAL LOSSES: vomiting, diarrhoea, burns, pancreatitis	<10 mmol/l
MODEST ECF VOLUME EXCESS (NO OEDEMA)	water intoxication (NB post-operative, TURP) Inappropriate ADH secretion, glucocorticoid deficiency, hypothyroidism, drugs (eg chlorpropamide, carbamazepine), pain, emotion. . .	>20 mmol/l
ECF VOLUME EXCESS (OEDEMA)	acute and chronic renal failure nephrotic syndrome, cirrhosis, cardiac failure	>20 mmol/l <10 mmol/l

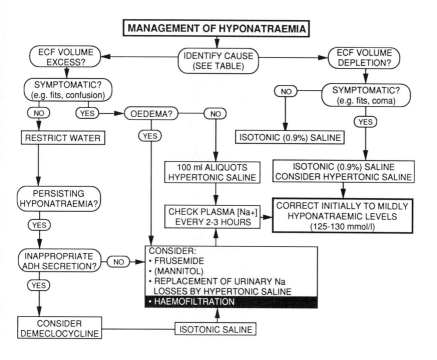

MANAGEMENT OF HYPONATRAEMIA

ECF VOLUME EXCESS? ← IDENTIFY CAUSE (SEE TABLE) → ECF VOLUME DEPLETION?

SYMPTOMATIC? (e.g. fits, confusion)
NO | YES → OEDEMA? ← NO

NO | SYMPTOMATIC? (e.g. fits, coma)
YES

RESTRICT WATER
YES

ISOTONIC (0.9%) SALINE

100 ml ALIQUOTS HYPERTONIC SALINE

ISOTONIC (0.9%) SALINE CONSIDER HYPERTONIC SALINE

PERSISTING HYPONATRAEMIA?
YES

CHECK PLASMA [Na+] EVERY 2-3 HOURS

CORRECT INITIALLY TO MILDLY HYPONATRAEMIC LEVELS (125-130 mmol/l)

INAPPROPRIATE ADH SECRETION? → NO →
YES

CONSIDER:
• FRUSEMIDE
• (MANNITOL)
• REPLACEMENT OF URINARY Na LOSSES BY HYPERTONIC SALINE
• HAEMOFILTRATION

CONSIDER DEMECLOCYCLINE

ISOTONIC SALINE

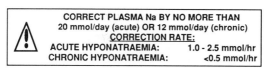

⚠ CORRECT PLASMA Na BY NO MORE THAN
20 mmol/day (acute) OR 12 mmol/day (chronic)
CORRECTION RATE:
ACUTE HYPONATRAEMIA: 1.0 - 2.5 mmol/hr
CHRONIC HYPONATRAEMIA: <0.5 mmol/hr

⚠ CARE WITH HYPERTONIC SOLUTIONS:
• n.b. ELDERLY, CARDIAC DYSFUNCTION
• CHECK Na LEVELS 2-3 HRLY
• CONSIDER PA CATHETER INSERTION

⚠ OVERDIURESIS CAN CAUSE HYPONATRAEMIA AND INTRAVASCULAR FLUID DEPLETION DESPITE PRESENCE OF OEDEMA

⚠ APPROPRIATE SPECIFIC THERAPY
e.g. • steroids for Addison's disease
• discontinue diuretic therapy
• discontinue drugs impairing water excretion

⚠ CORRECT COINCIDENT POTASSIUM AND MAGNESIUM LOSSES

9. Where haemodynamic instability is also present (e.g. in transurethral resection of the prostrate (TUR) syndrome) consider early insertion of a pulmonary artery catheter.
10. Spurious hyponatraemia may be due to (i) 'drip' artefacts if taken from same arm as a running infusion. (NB serum osmolality is not reduced by an appropriate extent) or (ii) analytical artefacts—hyperproteinaemia or hypertriglyceridaemia if using flame photometry of 'indirect' ion-specific electrodes (ISE). 'Direct-reading' ISEs are not susceptible to these changes in plasma water content though are sensitive to interference by excess heparin.

Hypernatraemia

1. Hypernatraemia is common during parenteral nutrition and usually relates to water deficiency.
2. It should be suspected in the obtunded patient.
3. Treat the cause. Aim for gradual correction of plasma Na (over 1–3 days) to avoid complications such as cerebral oedema.

Type	Aetiology	Urine
LOW TOTAL BODY Na	RENAL LOSSES: osmotic diuretics (glucose, urea, mannitol)	Na >20 mmol/l iso- or hypotonic
	EXTRA-RENAL LOSSES: excess sweating, diarrhoea in children	Na > 10 mmol/l hypertonic
NORMAL TOTAL BODY Na	RENAL LOSSES: nephrogenic diabetes insipidus central diabetes insipidus hypodipsia	Na variable hypo-, iso- or hypertonic
	EXTRA-RENAL LOSSES: respiratory and dermal insensible losses	Na variable hypertonic
INCREASED TOTAL BODY Na	Conn's syndrome, Cushing's syndrome, hypertonic sodium bicarbonate, sodium chloride tablets	Na >20 mmol/l iso- or hypertonic

Bibliography

Arieff AI. Hyponatremia, convulsions, respiratory arrest, and permanent brain damage after elective surgery in healthy women. N Engl J Med 1986; 314: 1529–35.

Ayus JC, Olivero JJ, and Frommer JP. Rapid correction of severe hyponatraemia with intravenous hypertonic saline solution. Am J Med 1982; 72: 43–48.

Berl T, Anderson RJ, McDonald KM, and Schrier RW. Clinical disorders of water metabolism. Kidney Int. 1976; 10: 117–126.

Cluitmans FHM and Meinders AE. Management of severe hyponatremia: Rapid or slow correction? Am J Med 1990; 88: 161–166.

Foote JW. Hyponatraemia: Diagnosis and management. Hosp Update 1990; 12: 248–258.

Larner AJ, Vickers CR, Adu D, Buchles JAC, Elias E and Neuberger J. Correction of severe hyponatraemia by continuous arteriovenous haemofiltration before liver transplantation. Br Med J 1988; 297: 1514–1515.

Laureno R and Karp BI. Pontine and extra-pontine myelinolysis following rapid correction of hyponatraemia. Lancet 1988 (i) 1439–1441.

Rossi NF. Crucial and practical aspects in the therapy of hyponatremia. Intensive & Crit Care Digest 1989; 8: 35–37.

Singer M, Patel M, Webb AR, and Bullen C. Transurethral syndrome— Time for re-appraisal? Crit Care Med 1990; 18: 1479–1480.

Sterns RH. Severe symptomatic hyponatraemia: Treatment and outcome. A study of 64 cases. Ann Intern Med 1987; 107: 656–664.

Sterns RH. The treatment of hyponatraemia: first, do no harm. Am J Med 1990; 88: 557–560.

Sterns RH, Riggs JE, and Schochet SS. Osmotic demyelination syndrome following correction of hyponatremia. N Engl J Med 1986; 314: 1535–1542.

Sterns RH, Thomas DJ, and Herndon RM Brain dehydration and neurologic deterioration after rapid correction of hyponatraemia. Kidney Int 1989; 35: 69–75.

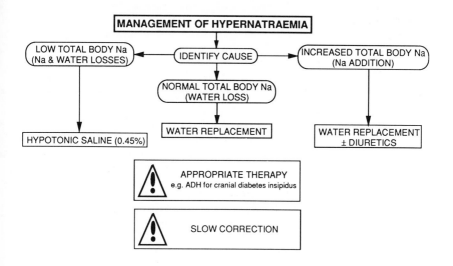

MANAGEMENT OF HYPERNATRAEMIA

LOW TOTAL BODY Na
(Na & WATER LOSSES)

IDENTIFY CAUSE

INCREASED TOTAL BODY Na
(Na ADDITION)

NORMAL TOTAL BODY Na
(WATER LOSS)

HYPOTONIC SALINE (0.45%)

WATER REPLACEMENT

WATER REPLACEMENT
± DIURETICS

⚠ APPROPRIATE THERAPY
e.g. ADH for cranial diabetes insipidus

⚠ SLOW CORRECTION

5.3: Metabolic acidosis

Treatment should always be directed at the underlying cause rather than manipulation of the pH.

Sodium bicarbonate therapy

1. The $PaCO_2$ may rise if minute volume is not increased. This may result in intracellular acidosis which can depress myocardial cell function.
2. The decrease in plasma ionized calcium may also cause a decrease in myocardial contractility. Significantly worse haemodynamic effects have been reported with bicarbonate compared to equimolar saline in patients with severe heart failure.
3. There is no human evidence that bicarbonate improves myocardial contractility or increases the response to circulating catecholamines in acidosis. Haemodynamic responses to sodium bicarbonate and equimolar sodium chloride solutions were shown to be similar in ICU patients with lactic acidosis; the hypertonicity of the bicarbonate solution appears more significant than its alkalinizing effect.
4. Bicarbonate may cause hyperosmolality, hypernatraemia, hypokalaemia and sodium overload.
5. Bicarbonate may cause arterial hypoxaemia and decreased tissue oxygenation.
6. Sodium bicarbonate does have a place in the management of acid retention or alkali loss, e.g. chronic renal failure, renal tubular acidosis, fistulae, etc. . . Replacement of fluid and potassium losses should be attempted first.

Miscellaneous

1. Though offering theoretical advantages, buffers such as Carbicarb and THAM (tri-hydroxy-methyl-aminomethane) have not been demonstrated superior to bicarbonate in the treatment of acidotic adults.
2. Type B lactic acidosis, i.e. lactaemia in the presence of adequate organ perfusion, may occur from a number of rare causes, e.g. phenformin therapy, fructose infusion, sorbitol, ethylene glycol. Aim treatment at discontinuing the causative factor and, possibly, using bicarbonate if the patient is very symptomatic.
3. The treatment of salicylate overdose remains controversial. Urinary alkalinization appears to be more important than a diuresis in salicylate elimination. Some advocate alkaline diuresis whereas others claim alkalinization of the urine alone is adequate for moderate poisoning (level 500–800mg/l). Haemodialysis is recommended for severe poisoning (>800 mg/l).
4. Alkaline diuresis is indicated in the treatment of myoglobinuria (to keep pH $\geqslant$5).

Bibliography

Arieff AI. Treatment of metabolic acidosis in low flow states: should we administer bicarbonate? In: Update in intensive care and emergency medicine No.8. Vincent J-L, Ed. Springer-Verlag, Berlin. 1989; 322–330.

Bersin RM and Arieff AI. Improved hemodynamic function during hypoxia with carbicarb, a new buffering agent for the management of acidosis. Circulation 1988; 77: 227–233.

Bersin RM, Chatterjee K, and Arieff AI. Metabolic and hemodynamic consequences of sodium bicarbonate administration in patients with heart disease. Am J Med 1989; 87: 5–6.

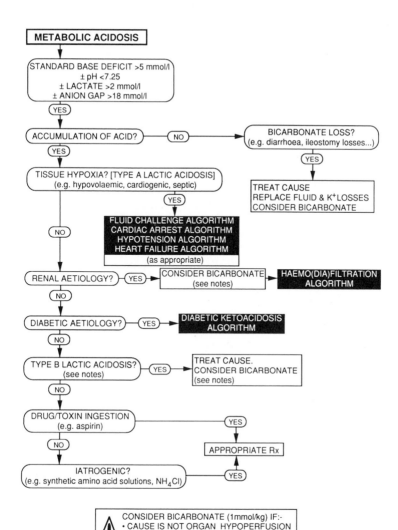

METABOLIC ACIDOSIS

STANDARD BASE DEFICIT >5 mmol/l
± pH <7.25
± LACTATE >2 mmol/l
± ANION GAP >18 mmol/l

YES

ACCUMULATION OF ACID? — NO — BICARBONATE LOSS?
(e.g. diarrhoea, ileostomy losses...)

YES — YES

TISSUE HYPOXIA? [TYPE A LACTIC ACIDOSIS]
(e.g. hypovolaemic, cardiogenic, septic)

TREAT CAUSE
REPLACE FLUID & K⁺LOSSES
CONSIDER BICARBONATE

YES

FLUID CHALLENGE ALGORITHM
CARDIAC ARREST ALGORITHM
HYPOTENSION ALGORITHM
HEART FAILURE ALGORITHM
(as appropriate)

NO

RENAL AETIOLOGY? — YES — CONSIDER BICARBONATE
(see notes) — HAEMO(DIA)FILTRATION
ALGORITHM

NO

DIABETIC AETIOLOGY? — YES — DIABETIC KETOACIDOSIS
ALGORITHM

NO

TYPE B LACTIC ACIDOSIS?
(see notes) — YES — TREAT CAUSE.
CONSIDER BICARBONATE
(see notes)

NO

DRUG/TOXIN INGESTION
(e.g. aspirin) — YES

NO — APPROPRIATE Rx

IATROGENIC?
(e.g. synthetic amino acid solutions, NH₄Cl) — YES

⚠ CONSIDER BICARBONATE (1mmol/kg) IF:-
• CAUSE IS NOT ORGAN HYPOPERFUSION
• NO ANURIA,OLIGURIA or FLUID OVERLOAD
• PROLONGED CARDIAC ARREST

Better OS and Stein JH. Early management of shock and prophylaxis of acute renal failure in traumatic rhabdomyolysis. N Engl J Med 1990; 322: 825–829.

Cooper DJ, Walley KR, Wiggs BR and Russell JA. Bicarbonate does not improve hemodynamics in critically ill patients who have lactic acidosis. Ann Intern Med 1990; 112: 492–498.

Graefe V, Miluntinovich J. Follette WC, Vizzo JE, Babb AL, and Scribner JH. Less dialysis induced morbidity and vascular instability with bicarbonate in dialysate. Ann Intern Med 1978; 88: 332–336.

Makisalo HJ, Soini HO, Nordin AJ, and Hockerstedt KAV. Effects of bicarbonate therapy on tissue oxygenation during resuscitation of hemorrhagic shock. Crit Care Med 1989; 17: 1170–1174.

Prestcott LF, Balali-Mood M, Critichley JAJH, Johnstone AF and Proudfoot AT. Diuresis of urinary alkalinzation for salicylate poisoning? Br Med J 1982; 285: 1383–1386.

Stacpoole PW. Lactic acidosis: the case against bicarbonate therapy. Ann Intern Med 1986; 105: 276–279.

5.4: Metabolic alkalosis

Treatment
1. Replacement of fluid, sodium, chloride and potassium losses are often sufficient to restore acid-base balance.
2. Active treatment is rarely necessary. If so, give ammonium chloride 5g tds p/o or hydrochloric acid given centrally as 1 N solution (1 mmol HCl/ml water) at a rate not exceeding 1 mmol/kg/hr.
3. Compensatory metabolic alkalosis for long-standing respiratory acidosis, followed by correction of the acidosis, for instance by mechanical ventilation, will lead to an uncompensated metabolic alkalosis. This will usually correct with time though treatments such as acetazolamide can be considered. Mechanical 'hypoventilation' i.e. maintaining hypercapnia, can also be considered.

Miscellaneous
1. Urinary pH is often acidic in the presence of a significant metabolic alkalosis.
2. Alkalosis is associated with an increased alveolar-arterial oxygen difference and hypoxaemia.
3. Correction of alkalosis causes an increased PaO_2 related to the Bohr effect and to enhanced hypoxic pulmonary vasoconstriction. An unwanted, though possibly transient, effect may be increases in pulmonary vascular pressures.

Bibliography
Brimioulle S, Vincent J-L, Dufaye P, Berre J, Degaute JP, and Kahn RJ. Hydrochloric acid infusion for treatment of metabolic alkalosis effects on acid-base balance and oxygenation. Crit Care Med 1985; 13: 738–742.
Brimioulle S, Kahn RJ. Effects of metabolic alkalosis on pulmonary gas exchange. Am Rev Respir Dis 1990; 141: 1185–1189.
Parsons V. Acute metabolic emergencies. In: Introduction to Intensive Care. Potter D, ed. Farrand Press, London 1990, pp 248–250.
Worthley LIG. Acid-base balance and disorders. In: Intensive Care Manual, Oh TE, Ed. Butterworths, Sydney 1990, pp 500–501.

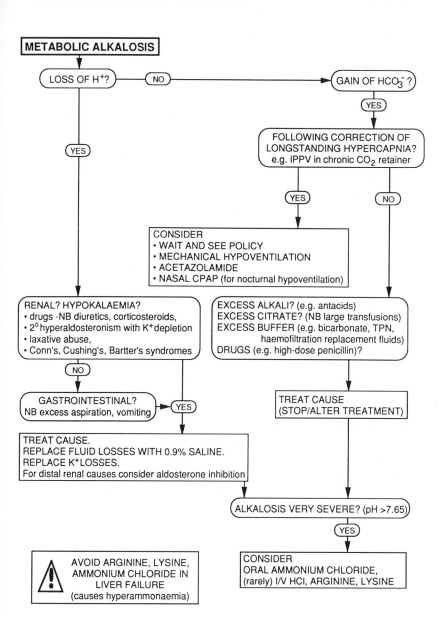

METABOLIC ALKALOSIS

LOSS OF H⁺? ── NO ──────────────→ GAIN OF HCO₃⁻?

GAIN OF HCO₃⁻? — YES

FOLLOWING CORRECTION OF
LONGSTANDING HYPERCAPNIA?
e.g. IPPV in chronic CO_2 retainer

— YES —   — NO —

CONSIDER
• WAIT AND SEE POLICY
• MECHANICAL HYPOVENTILATION
• ACETAZOLAMIDE
• NASAL CPAP (for nocturnal hypoventilation)

LOSS OF H⁺? — YES

RENAL? HYPOKALAEMIA?
• drugs -NB diuretics, corticosteroids,
• 2⁰ hyperaldosteronism with K⁺depletion
• laxative abuse,
• Conn's, Cushing's, Bartter's syndromes

EXCESS ALKALI? (e.g. antacids)
EXCESS CITRATE? (NB large transfusions)
EXCESS BUFFER (e.g. bicarbonate, TPN,
              haemofiltration replacement fluids)
DRUGS (e.g. high-dose penicillin)?

— NO —

GASTROINTESTINAL?
NB excess aspiration, vomiting ── YES

TREAT CAUSE
(STOP/ALTER TREATMENT)

TREAT CAUSE.
REPLACE FLUID LOSSES WITH 0.9% SALINE.
REPLACE K⁺LOSSES.
For distal renal causes consider aldosterone inhibition

ALKALOSIS VERY SEVERE? (pH >7.65)

— YES

⚠ AVOID ARGININE, LYSINE,
AMMONIUM CHLORIDE IN
LIVER FAILURE
(causes hyperammonaemia)

CONSIDER
ORAL AMMONIUM CHLORIDE,
(rarely) I/V HCl, ARGININE, LYSINE

5.5: Hyperglycaemic ketoacidosis

Fluid management
1. In severe diabetic ketoacidosis (DKA) approximate losses are:
   - water 5–10 litres
   - sodium 400–700 mmol
   - potassium 250–700 mmol.
2. The fluid loss is relatively hypotonic (urine sodium concentration is about 55 mmol/l) as about half is derived from the intracellular compartment.
3. Fluid and electrolyte repletion should not follow a strict regimen but should be tailored to individual needs.
4. Fluid replacement with 0.9% saline can be used though caution should be exercised to prevent sodium overload. 5% glucose should be substituted after replacing the sodium debt.
5. Hypotonic (0.45%) saline resuscitation is appropriate in the non-shocked patient but in shock it is important to restore the circulating blood volume as quickly as possible with colloid solutions.
6. Over-rapid rehydration may result in cerebral oedema. Patients with impaired cardiac or renal function may also be compromised.

Management of hyperglycaemia
1. Hyperglycaemia should be corrected gradually at a rate of 2–4 mmol/hr.
2. Insulin infusion should be continued, even after achieving normoglycaemia, as there may be insufficient circulating endogenous insulin to enable glucose to enter the cells. This may require additional glucose, possibly at a higher concentration.
   NB 5% glucose only contains 20 Cal/100 ml.

Other aspects
1. A nasogastric tube should always be inserted as gastric emptying is often delayed and acute gastric dilatation is common; the risk of aspiration is increased.
2. Bicarbonate should not be used routinely, even for severe acidosis. It causes:
   - Increased intracellular acidosis.
   - Depressed respiration due to relative CSF alkalosis.
3. Antibiotics should only be given for proven infection or where there is a high index of clinical suspicion.
4. Abdominal pain should not be dismissed as part of the syndrome.
5. Magnesium and calcium deficiencies should be treated (e.g. $MgSO_4$ 20mmol over 2–3 hr).

Hyperosmolar hyperglycaemic non-ketotic states
1. More common in elderly, non-insulin dependent diabetics.
2. Fluid depletion is greater, coma more frequent and mortality much higher than in DKA.
3. Hyperosmolality may predispose thrombotic events. Unless otherwise contraindicated, these patients should be fully heparinized until full recovery (which may take 5+ days).
4. Hypotonic saline resuscitation should be slow (over 48–72 hr).
5. Patients may be hypersensitive to insulin and require lower doses.

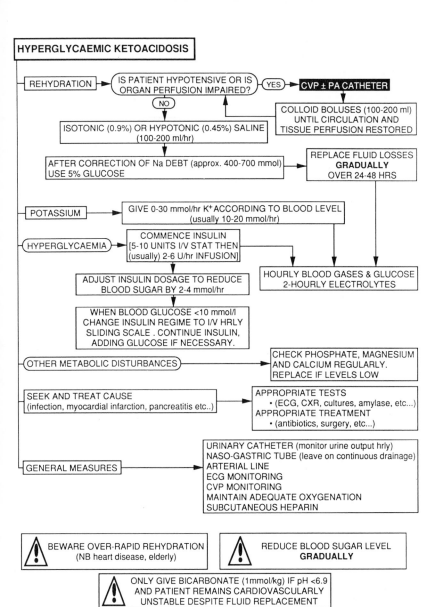

HYPERGLYCAEMIC KETOACIDOSIS

REHYDRATION → IS PATIENT HYPOTENSIVE OR IS ORGAN PERFUSION IMPAIRED? → YES → **CVP ± PA CATHETER**

NO

COLLOID BOLUSES (100-200 ml) UNTIL CIRCULATION AND TISSUE PERFUSION RESTORED

ISOTONIC (0.9%) OR HYPOTONIC (0.45%) SALINE (100-200 ml/hr)

AFTER CORRECTION OF Na DEBT (approx. 400-700 mmol) USE 5% GLUCOSE

REPLACE FLUID LOSSES **GRADUALLY** OVER 24-48 HRS

POTASSIUM → GIVE 0-30 mmol/hr K⁺ ACCORDING TO BLOOD LEVEL (usually 10-20 mmol/hr)

HYPERGLYCAEMIA → COMMENCE INSULIN [5-10 UNITS I/V STAT THEN (usually) 2-6 U/hr INFUSION]

ADJUST INSULIN DOSAGE TO REDUCE BLOOD SUGAR BY 2-4 mmol/hr

HOURLY BLOOD GASES & GLUCOSE 2-HOURLY ELECTROLYTES

WHEN BLOOD GLUCOSE <10 mmol/l CHANGE INSULIN REGIME TO I/V HRLY SLIDING SCALE. CONTINUE INSULIN, ADDING GLUCOSE IF NECESSARY.

OTHER METABOLIC DISTURBANCES → CHECK PHOSPHATE, MAGNESIUM AND CALCIUM REGULARLY. REPLACE IF LEVELS LOW

SEEK AND TREAT CAUSE (infection, myocardial infarction, pancreatitis etc..)

APPROPRIATE TESTS
• (ECG, CXR, cultures, amylase, etc...)
APPROPRIATE TREATMENT
• (antibiotics, surgery, etc...)

GENERAL MEASURES → URINARY CATHETER (monitor urine output hrly) NASO-GASTRIC TUBE (leave on continuous drainage) ARTERIAL LINE ECG MONITORING CVP MONITORING MAINTAIN ADEQUATE OXYGENATION SUBCUTANEOUS HEPARIN

⚠ BEWARE OVER-RAPID REHYDRATION (NB heart disease, elderly)

⚠ REDUCE BLOOD SUGAR LEVEL **GRADUALLY**

⚠ ONLY GIVE BICARBONATE (1mmol/kg) IF pH <6.9 AND PATIENT REMAINS CARDIOVASCULARLY UNSTABLE DESPITE FLUID REPLACEMENT

Bibliography

Alberti KGMM and Hockaday TDR. Diabetes mellitus. In: Weatherall DJ, Ledingham JG, Warrell DA, eds. Oxford Textbook of Medicine. Oxford; Oxford University Press. 1987; 9.51–9. 101.

Arieff AL and Kleeman CR. Studies on mechanisms of cerebral edema in diabetic comas: effects of hyperglycaemia and rapid lowering of plasma glucose in normal rabbits. J Clin Invest 1973; 52: 571–583.

Hillman K. Fluid resuscitation in diabetic emergencies—a reappraisal. Intensive Care Med 1987; 13: 4–8.

Richardson JE and Donaldson MDC. Diabetic emergencies. Prescribers Journal 1989; 29: 174–182.

6. Gastrointestinal

6.1: Acute gastrointestinal bleeding

Stress ulceration
1. Virtually all critically-ill patients show endoscopic evidence of stress ulceration.
2. The current incidence of significant bleeding from stress ulceration is <10–20% with a direct mortality less than 1–2%, whether the patient is receiving prophylaxis or not.
3. H_2 antagonists ± antacid to keep pH ⩾5 are usually effective in stopping bleeding from stress ulceration once it occurs. Surgery is very rarely necessary.
4. Prophylaxis of stress ulceration should concentrate on general measures such as prevention of hypotension, hypoxia, coagulopathy, etc. . . H_2 antagonists, sucralfate and antacids are probably best reserved for 'high-risk' patients.
5. The best prophylactic agent is still open to question. Ranitidine is easy to administer by bolus injection although gastric pH should be monitored to ensure adequate prophylaxis (pH ⩾4).This does not need to be monitored with sucralfate which is administered orally or via a nasogastric tube. The evidence for sucralfate reducing the incidence of nosocomial pneumonia is conflicting.

Management
1. CVP ± PA catheter monitoring may be useful, especially in the elderly and those with cardiac dysfunction. Caution should be exercised as clotting abnormalities are often present.
2. Colloid is a better volume expander than blood. Shocked patients should receive a combination, particularly as most stored blood is plasma deplete.
3. Coagulopathies should be actively corrected.
4. Platelet transfusions should be given—despite normal levels—if aspirin has been ingested recently.
5. H_2 antagonists and antacids will not generally stop bleeding from non-stress ulcers though they do facilitate ulcer healing.
6. Unless varices are present, a NG tube should be inserted to give medication and assess the rate of bleeding.

Diagnosis
1. The patient should be adequately resuscitated first.
2. Early endoscopy should be performed for all large bleeds. Therapeutic injection, laser therapy, etc. . . may prove highly successful.
3. >50% of cases of upper GI bleeding in cirrhotics have a non-variceal origin. Dual pathology should also be considered.

Surgery
1. Surgeons should remain well-informed throughout.
2. Mortality is significantly higher in massive or recurrent bleeding. Early surgery should be considered in these cases.
3. Early surgery should be considered in the elderly.

Varices
1. Early endoscopy ± sclerotherapy should be performed for all variceal bleeds.

2. A Sengstaken-type tube should be inserted for large variceal bleeds.
3. The oesophageal balloon should not be inflated. Compression of the gastric balloon on the cardia (and fundus) should control bleeding satisfactorily. Adequate traction of the tube should be maintained (approx. 1 kg).
4. Continuous aspiration of the oesophageal lumen of tube reduces the risk of aspiration.
5. Vasopressin will only control bleeding temporarily in 60% of cases. Simultaneous administration of GTN may reduce side-effects.

Lower GI bleeding

1. Endoscopy, isotope scanning and angiography may delineate the site of the lesion. Angiography will only reveal a lesion if there is active bleeding of at least 0.5 ml/min.
2. Early surgery is the preferred treatment for massive lower GI bleeding.

Bibliography

Boley SJ, Brandt LJ, and Frank MS. Severe lower intestinal bleeding: Diagnosis and treatment. Clin Gastroenterol. 1981; 10: 65–91.

Chung SC, Leung JW, Steele RJ, and Lia K. Endoscopic injection of adrenaline for actively bleeding ulcers. Br Med J 1988; 296: 1631–1633.

Gimson AE, Westaby D, Hegarty J, et al. A randomized trial of vasopressin and vasopressin plus nitroglycerine in the control of acute variceal haemorrhage. Hepatology 1986; 6: 410–413.

Lacroix J, Infante-Rivard C, Jenicek M, and Gauthier M. Prophylaxis of upper gastrointestinal bleeding in intensive care units. A meta-analysis. Crit Care Med 1989; 17: 862–869.

Laurence BH. Acute gastrointestinal bleeding. In: Intensive Care Manual 3rd Edition. Oh TE. (Ed.) Butterworths Sydney 1990 pp 219–224.

Laurence BH and Cotton PB. Bleeding gastroduodenal ulcers: Non-operative management. World J Surg 1987; 11: 295–303.

Leicester RJ. Gastrointestinal haemorrhage. Medicine International 1990; 77: 3188–3193.

Reusser P, Gyr K, Scheidegger D, Buchmann B, Buser M, and Zimmerli W. Prospective endoscopic study of stress erosions and ulcers in critically ill neurosurgical patients: Current incidence and effect of acid-reducing prophylaxis. Crit Care Med 1990 18: 270–274.

Wilcox CM and Spenney JG. Stress ulcer prophylaxis in medical patients: who, what and how much? Am J Gastroenterol 1988; 83: 1199–1210.

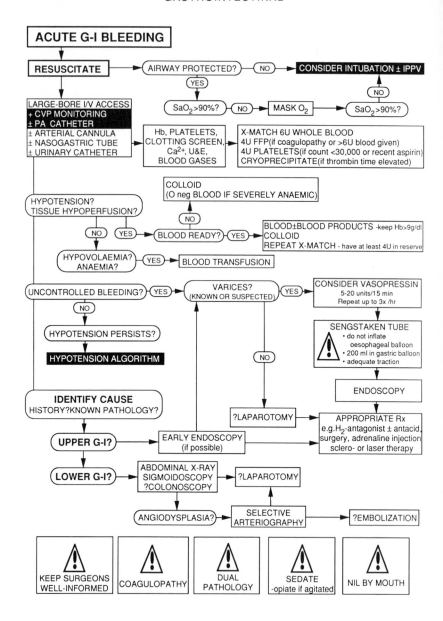

ACUTE G-I BLEEDING

RESUSCITATE → AIRWAY PROTECTED? → NO → **CONSIDER INTUBATION ± IPPV**

YES

SaO$_2$ >90%? → NO → MASK O$_2$ → SaO$_2$ >90%?

NO

LARGE-BORE I/V ACCESS
+ CVP MONITORING
± PA CATHETER
± ARTERIAL CANNULA
± NASOGASTRIC TUBE
± URINARY CATHETER

Hb, PLATELETS,
CLOTTING SCREEN,
Ca^{2+}, U&E,
BLOOD GASES

X-MATCH 6U WHOLE BLOOD
4U FFP(if coagulopathy or >6U blood given)
4U PLATELETS(if count <30,000 or recent aspirin)
CRYOPRECIPITATE(if thrombin time elevated)

COLLOID
(O neg BLOOD IF SEVERELY ANAEMIC)

HYPOTENSION?
TISSUE HYPOPERFUSION? → NO / YES

NO

BLOOD READY? → YES → BLOOD±BLOOD PRODUCTS -keep Hb>9g/dl
COLLOID
REPEAT X-MATCH - have at least 4U in reserve

HYPOVOLAEMIA?
ANAEMIA? → YES → BLOOD TRANSFUSION

UNCONTROLLED BLEEDING? → YES → VARICES?
(KNOWN OR SUSPECTED) → YES → CONSIDER VASOPRESSIN
5-20 units/15 min
Repeat up to 3x /hr

NO

HYPOTENSION PERSISTS?

HYPOTENSION ALGORITHM

NO

SENGSTAKEN TUBE
⚠ • do not inflate
oesophageal balloon
• 200 ml in gastric balloon
• adequate traction

ENDOSCOPY

IDENTIFY CAUSE
HISTORY?KNOWN PATHOLOGY?

?LAPAROTOMY

APPROPRIATE Rx
e.g.H$_2$-antagonist ± antacid,
surgery, adrenaline injection,
sclero- or laser therapy

UPPER G-I? → EARLY ENDOSCOPY
(if possible)

LOWER G-I? → ABDOMINAL X-RAY
SIGMOIDOSCOPY
?COLONOSCOPY → ?LAPAROTOMY

ANGIODYSPLASIA? → SELECTIVE
ARTERIOGRAPHY → ?EMBOLIZATION

⚠ KEEP SURGEONS
WELL-INFORMED

⚠ COAGULOPATHY

⚠ DUAL
PATHOLOGY

⚠ SEDATE
-opiate if agitated

⚠ NIL BY MOUTH

Notes

6.2/6.3: Fulminant hepatic failure/Acute on chronic hepatic failure

Aetiology
1. Viruses and paracetamol are the commonest causes of acute liver failure.
2. N-acetyl cysteine appears effective up to 36 hr after paracetamol ingestion.

Encephalopathy and cerebral oedema
1. Early administration of lactulose is beneficial in early encephalopathy.
2. Mortality is much higher for Grades III–IV.
3. Cerebral oedema is much commoner for Grades III–IV (75% with Grade IV). A subdural transducer should ideally be used to maintain cerebral perfusion pressure [CPP] (= mean arterial pressure [MAP]—intracranial pressure [ICP] >50 mmHg and ICP <20–25 mmHg.
4. Avoid hypotension, hypoxaemia, and hypercapnia.
5. Mannitol can be given hourly if necessary to decrease ICP and increase CPP. Aim to remove via urine (or ultrafiltrate) 3× volume of mannitol given.
6. If mannitol proves ineffective give thiopentone (125–250 mg over 15 min then 50–250 mg/hr for 4 hr only).
7. Hyperventilation is useful for short periods only.
8. Steroids are ineffective.

Renal failure
1. Occurs in 30–75% of Grade IV encephalopathy. Commoner with paracetamol.
2. Consider hypovolaemia, hypotension and nephrotoxic drugs.
3. Do not give high dose frusemide. Take care with huge dose mannitol.
4. For renal replacement heparin is usually needed in higher doses due to reduced Antithrombin III. Prostacyclin is probably better.
5. With renal replacement (especially dialysis) care needed to avoid hypovolaemia or too rapid removal of solute—risk of cerebral oedema.

Metabolic
1. Hypoglycaemia is very common. Monitor hourly. Give 50% glucose if necessary.
2. Both metabolic alkalosis and acidosis can occur. Correct hypokalaemia.

Cardiorespiratory
1. Arrhythmias common. Correct obvious causes eg hypokalaemia.
2. Hypoxaemia common—V/Q mismatch due to shunting, infection, aspiration, etc. . .

Sepsis
1. The patient is immunocompromised.
2. Daily cultures are essential.
3. Consider antibiotics and antifungals in recovering patients in whom improvement in prothrombin time ceases.

Transplantation
1. Early discussion with specialist/specialist centre.
2. Important factors for consideration: age, aetiology, interval between jaundice and encephalopathy, bilirubin (>300 μmol/l) and prothrombin time (>50 sec).

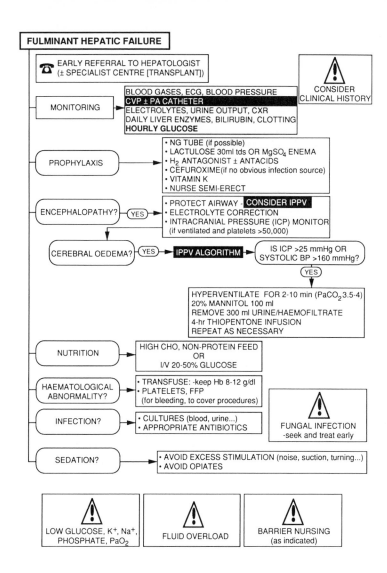

FULMINANT HEPATIC FAILURE

EARLY REFERRAL TO HEPATOLOGIST (± SPECIALIST CENTRE [TRANSPLANT])

CONSIDER CLINICAL HISTORY

MONITORING
BLOOD GASES, ECG, BLOOD PRESSURE
CVP ± PA CATHETER
ELECTROLYTES, URINE OUTPUT, CXR
DAILY LIVER ENZYMES, BILIRUBIN, CLOTTING
HOURLY GLUCOSE

PROPHYLAXIS
• NG TUBE (if possible)
• LACTULOSE 30ml tds OR MgSO$_4$ ENEMA
• H$_2$ ANTAGONIST ± ANTACIDS
• CEFUROXIME(if no obvious infection source)
• VITAMIN K
• NURSE SEMI-ERECT

ENCEPHALOPATHY? (YES)
• PROTECT AIRWAY - CONSIDER IPPV
• ELECTROLYTE CORRECTION
• INTRACRANIAL PRESSURE (ICP) MONITOR (if ventilated and platelets >50,000)

CEREBRAL OEDEMA? (YES) → IPPV ALGORITHM
IS ICP >25 mmHg OR SYSTOLIC BP >160 mmHg? (YES)

HYPERVENTILATE FOR 2-10 min (PaCO$_2$ 3.5-4)
20% MANNITOL 100 ml
REMOVE 300 ml URINE/HAEMOFILTRATE
4-hr THIOPENTONE INFUSION
REPEAT AS NECESSARY

NUTRITION
HIGH CHO, NON-PROTEIN FEED
OR
I/V 20-50% GLUCOSE

HAEMATOLOGICAL ABNORMALITY?
• TRANSFUSE: -keep Hb 8-12 g/dl
• PLATELETS, FFP (for bleeding, to cover procedures)

INFECTION?
• CULTURES (blood, urine...)
• APPROPRIATE ANTIBIOTICS
FUNGAL INFECTION -seek and treat early

SEDATION?
• AVOID EXCESS STIMULATION (noise, suction, turning...)
• AVOID OPIATES

LOW GLUCOSE, K$^+$, Na$^+$, PHOSPHATE, PaO$_2$

FLUID OVERLOAD

BARRIER NURSING (as indicated)

Acute on chronic liver failure

1. Ascites is far commoner. Infected ascites should be ruled out as cause of decompensation.
2. Other causes of G–I bleeding other than varices e.g. peptic ulcer should be considered.
3. Neomycin may be useful for a few days only to reduce breakdown of gastrointestinal blood.

Bibliography

Harrison PM, Keays R, Bray GP, Alexander GJM, and Williams R. Improved outcome of paracetamol-induced fulminant hepatic failure by late administration of acetylcysteine. Lancet 1990; i: 1572–1573.
Stoller JK. As the liver goes, so goes the lung. Chest 1990; 97: 1028–1030.
Williams R. Acute liver failure: an overview. In: Update in Intensive Care & Emergency Medicine No.8. Ed. Vincent J-L. Springer-Verlag, Berlin 1989 pp. 376–385.

ACUTE ON CHRONIC HEPATIC FAILURE

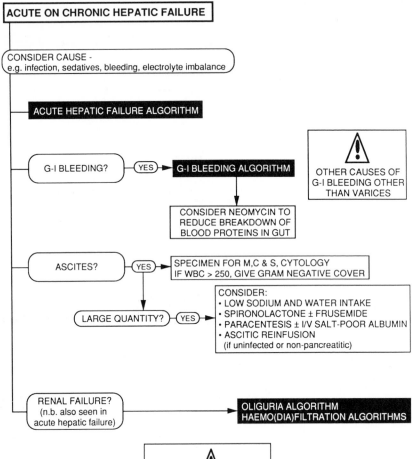

CONSIDER CAUSE -
e.g. infection, sedatives, bleeding, electrolyte imbalance

ACUTE HEPATIC FAILURE ALGORITHM

G-I BLEEDING? — YES → **G-I BLEEDING ALGORITHM**

⚠ OTHER CAUSES OF G-I BLEEDING OTHER THAN VARICES

CONSIDER NEOMYCIN TO REDUCE BREAKDOWN OF BLOOD PROTEINS IN GUT

ASCITES? — YES → SPECIMEN FOR M,C & S, CYTOLOGY
IF WBC > 250, GIVE GRAM NEGATIVE COVER

LARGE QUANTITY? — YES → CONSIDER:
• LOW SODIUM AND WATER INTAKE
• SPIRONOLACTONE ± FRUSEMIDE
• PARACENTESIS ± I/V SALT-POOR ALBUMIN
• ASCITIC REINFUSION
  (if uninfected or non-pancreatitic)

RENAL FAILURE?
(n.b. also seen in acute hepatic failure) → **OLIGURIA ALGORITHM**
HAEMO(DIA)FILTRATION ALGORITHMS

⚠ AVOID HYPOVOLAEMIA, FLUID OVERLOAD AND ELECTROLYTE IMBALANCE. MONITOR CREATININE

Notes

7. Neurological

7.1: Head injury

CPP (cerebral perfusion pressure) = Mean Arterial Pressure − Intracranial Pressure (ICP)

Management of raised intracranial pressure
1. Significant increases in ICP occur in half the patients with severe head injury during the first 72 hr.
2. Hyperventilation is the mainstay of therapy for raised ICP but is only effective for short periods.
3. Severe alkalosis should be avoided since it increases cerebral vascular resistance and promotes cerebral ischaemia.
4. CSF bicarbonate adjusts to low values of $PaCO_2$ within 4–6 hr. Thereafter, rapid normalisation of $PaCO_2$ will lead to increases in ICP.
5. Fluid restriction and active diuresis (mannitol and/or frusemide) will usefully reduce ICP. This must not be done at the expense of destabilizing cardiovascular status. A plasma osmolality of 310–320 mOsm/kg should not be exceeded.
6. Hyperventilation and dehydration therapy will satisfactorily control 80% of cases of raised ICP. Barbiturate therapy will be successful in half of the resistant cases though inotrope/pressor therapy may have to be added.
7. Sedation, analgesia and paralysis will reduce ICP. The need for sedation usually outweighs the problems associated with neurological assessment. Coughing, noise, purposeless agitation and pain should be avoided. Lignocaine and pre-oxygenation may be useful pre-endotracheal suction.
8. Steroids not demonstrated to be beneficial in head injury though may be useful in spinal injury if started within 8 hr (methylprednisolone 30 mg/kg bolus then 5.4 mg/kg/hr for 23 hr).
9. There is a variable response of ICP to PEEP.

Blood pressure management
1. CPP should be maintained over 50–60 mmHg. In the absence of ICP monitoring, MAP should be maintained >80 mmHg.
2. Hypertension should not generally be treated. However, if treatment is contemplated:
   - only treat if cerebral perfusion pressure (CPP = MAP − ICP) is also raised.
   - monitor ICP (if possible) to ensure CPP is maintained.
   - avoid direct vasodilators which will increase cerebral blood volume and therefore ICP.
   - use β blockers to reduce myocardial effects of excessive circulating catecholamines.
3. Pressor therapy may be indicated if the patient is relatively hypotensive.

ICP monitoring
1. Indications
   - GCS ≤ 8 and any abnormality on CT scan.
   - GCS ≤ 8 and a normal CT scan but any two of the following:
     - age > 40 years
     - hypotension
     - decerebrate posturing.
   - GCS > 8 but:

- requiring general anaesthesia for treatment of other injuries
- requiring treatment that is likely to increase ICP, e.g. high levels of PEEP.

2. Contraindications:
   - coagulopathy.
3. Confirmation of satisfactory ICP placement:
   - pulsation in phase with blood pressure
   - increases with transient jugular venous compression or Valsalva manoeuvre.
4. ICP monitoring should be continued:
   - as long as ICP is elevated
   - during active management of ICP
   - for up to 3/7 in the absence of significant elevation.

Prophylactic antibiotics

1. Benzylpenicillin 1.2g I/V 6 hourly should be given in the following cases
   - Basal fracture.
   - Compound vault fracture.
   - Suspected or proven meningitis.

Seizures

1. Control of seizures
   - prevents further hypoxaemic cerebral damage
   - reduces cerebral oxygen requirement
   - reduces raised ICP.
2. Should be controlled with phenytoin 1g I/V given at a rate not exceeding 50mg/min then 300 mg daily. Plasma levels should be checked.
3. Surgically remediable causes should be sought.
4. Hypoglycaemia and other metabolic disturbances should be excluded.
5. Uncontrolled fits may necessitate a thiopentone infusion.

Miscellaneous

1. Treat hypo- or hyperglycaemia.
2. Correct metabolic disturbances.
3. Treat diabetes insipidus with I/V or I/M DDAVP 1–4 mg/day (NB this carries a grave prognosis).
4. Head injury induces a hypermetabolic state. Institute early enteral (or parenteral) nutrition.

Bibliography

Advanced Trauma Life Support Student Manual. Chapter 6: Head trauma. American College of Surgeons, Illinois 1989.

Borel C, Hanley D, Diringer MN, and Rogers MC. Intensive management of severe head injury. Chest 1990; 98: 180–189.

Bracken MB, Shepard MJ, and Collins WF *et al*. A randomised, controlled trial of methylprednisolone or naloxone in the treatment of acute spinal cord injury. Results of the Second National Acute Spinal Cord Injury Study. N Engl J Med 1990; 322: 1405–1411.

Bullock R and Teasdale G. Head injuries part 1 Br Med J 1990; 300: 1515–1518.

Bullock R and Teasdale G. Head injuries part 2 Br Med J 1990; 300: 1577–1579.

Dearden NM, Gibson JS, McDowall DG, Gibson RM, and Cameron MM. Effect of high-dose dexamethasone on outcome from severe head injury. J Neurosurg 1986; 64: 81–88.

Obrist WD, Langfitt TW, Jaggi JL, Cruz J, and Genarelli TA. Cerebral blood flow and metabolism in comatose patients with acute head injury: relationship with intracranial hypertension. J Neurosurg 1984; 61: 241–253.

Price D. Principles of managing head injuries. Care of the Critically Ill 1985; 1: 3–5.

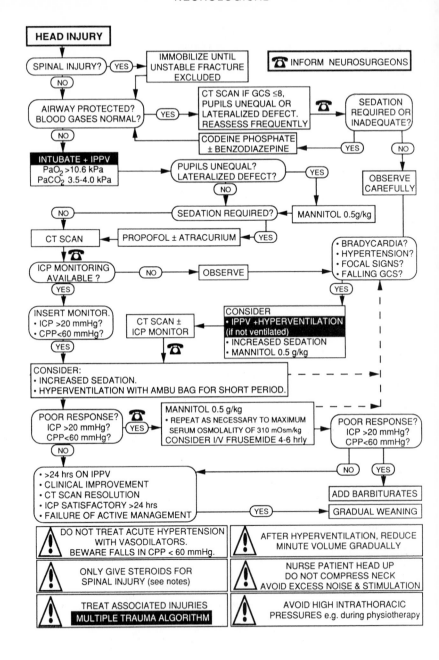

HEAD INJURY

SPINAL INJURY? —YES→ IMMOBILIZE UNTIL UNSTABLE FRACTURE EXCLUDED

☎ INFORM NEUROSURGEONS

NO

AIRWAY PROTECTED? BLOOD GASES NORMAL? —YES→ CT SCAN IF GCS ≤8, PUPILS UNEQUAL OR LATERALIZED DEFECT. REASSESS FREQUENTLY ☎ → SEDATION REQUIRED OR INADEQUATE?

NO

CODEINE PHOSPHATE ± BENZODIAZEPINE ←YES—

INTUBATE + IPPV
PaO$_2$ >10.6 kPa
PaCO$_2$ 3.5-4.0 kPa

PUPILS UNEQUAL? LATERALIZED DEFECT? —YES→

NO

OBSERVE CAREFULLY

NO ← SEDATION REQUIRED? ← MANNITOL 0.5g/kg

CT SCAN ← PROPOFOL ± ATRACURIUM —YES→

☎

ICP MONITORING AVAILABLE? —NO→ OBSERVE

• BRADYCARDIA?
• HYPERTENSION?
• FOCAL SIGNS?
• FALLING GCS?

YES

YES

INSERT MONITOR.
• ICP >20 mmHg?
• CPP<60 mmHg?

CT SCAN ± ICP MONITOR

CONSIDER
• IPPV +HYPERVENTILATION (if not ventilated)
• INCREASED SEDATION
• MANNITOL 0.5 g/kg

YES

☎

CONSIDER:
• INCREASED SEDATION.
• HYPERVENTILATION WITH AMBU BAG FOR SHORT PERIOD.

POOR RESPONSE? ☎
ICP >20 mmHg?
CPP<60 mmHg? —YES→ MANNITOL 0.5 g/kg
• REPEAT AS NECESSARY TO MAXIMUM SERUM OSMOLALITY OF 310 mOsm/kg
CONSIDER I/V FRUSEMIDE 4-6 hrly

POOR RESPONSE?
ICP >20 mmHg?
CPP<60 mmHg?

NO

NO   YES

• >24 hrs ON IPPV
• CLINICAL IMPROVEMENT
• CT SCAN RESOLUTION
• ICP SATISFACTORY >24 hrs
• FAILURE OF ACTIVE MANAGEMENT —YES→

ADD BARBITURATES

GRADUAL WEANING

⚠ DO NOT TREAT ACUTE HYPERTENSION WITH VASODILATORS. BEWARE FALLS IN CPP < 60 mmHg.

⚠ AFTER HYPERVENTILATION, REDUCE MINUTE VOLUME GRADUALLY

⚠ ONLY GIVE STEROIDS FOR SPINAL INJURY (see notes)

⚠ NURSE PATIENT HEAD UP DO NOT COMPRESS NECK AVOID EXCESS NOISE & STIMULATION

⚠ TREAT ASSOCIATED INJURIES MULTIPLE TRAUMA ALGORITHM

⚠ AVOID HIGH INTRATHORACIC PRESSURES e.g. during physiotherapy

Notes

7.2: Generalized seizures

1. Always consider underlying disorders such as space-occupying lesions, metabolic disturbances, etc. . . which require specific treatments.
2. EEG monitoring such as CFAM (cerebral function analysis monitoring) may be useful, especially in the ventilated patient.

Drug therapy
1. Monitor ECG when giving intravenous phenytoin. Give at a rate not exceeding 50 mg/min.
2. Both benzodiazepines and chlormethiazole may cause respiratory depression and hypoxaemia.
3. If large doses of chlormethiazole are used care should be taken as large volumes of an electrolyte-free solution are being infused.
4. High dose dexamethasone may be used if the patient has a known tumour, arteritis or parasitic CNS infection.
5. In alcoholics thiamine 100 mg I/V should be given.
6. Clonazepam should be used for myoclonic seizures.
7. $MgSO_4$ can be useful for intractable seizures. Loading dose 4 g (16 mmol) over 20 min. Maintenance dose 1–2 g/hr. NB maintain good urine output and monitor neuromuscular function if possible.
8. Treat all diagnoses of encephalitis as for herpes simplex virus with acyclovir 10 mg/kg 8 hourly for 10 days before diagnosis confirmed.
9. Sodium valproate can cause severe or fatal liver disfunction, particularly in combination with other anti-epileptic agents.
10. Phenytoin should not be given orally if patient is on enteral nutrition.

Supportive treatment
1. Circulation:
   - correct hypoxaemia
   - correct hypovolaemia and electrolyte disturbances (NB diabetes insipidus).
2. Generalized seizures may result in hyperthermia which can contribute to cerebral damage.
3. Prevent and treat cerebral oedema:
   - generous sedation
   - hypothermia
   - controlled hyperventilation
   - osmotic diuretics if indicated.

Bibliography
Brodie MJ. Status epilepticus in adults. Lancet 1990; 336: 551–552.
Dodson WE, Prenski AL, DeVivo DC, Goldring S, and Dodge PR. Management of seizure disorders: Selected asects. J Pediatr 1976; 89: 527–540.
Meldrum BS and Brierley JB. Prolonged epileptic seizures in primates. Ischemic cell change and its relation to ictal physiological events. Arch Neurol 1973; 28: 10–17.
Pritchard JA. Management of pre-eclampsia and eclampsia. Kidney Int 18; 259: 1980.
Trubuhovich RV. Management of severe or intractable convulsions including eclampsia. Int Anesthesiol Clin; 17: 201–238.
Yeoman P, Hutchinson A, Byrne A, Smith J, and Durham S. Etomidate infusion for the control of refractory status epilepticus. Intensive Care Med 1989; 15: 255–260.

GENERALIZED SEIZURES

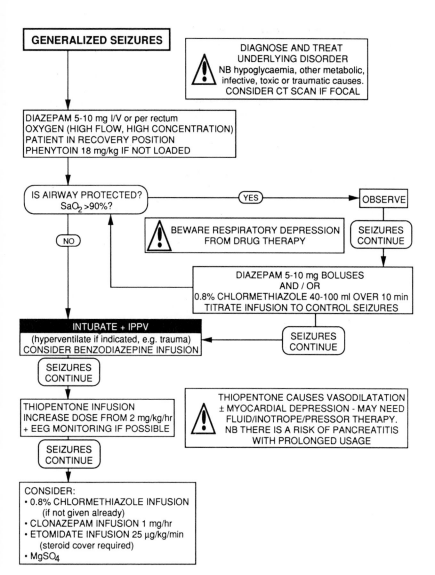

⚠ DIAGNOSE AND TREAT UNDERLYING DISORDER
NB hypoglycaemia, other metabolic, infective, toxic or traumatic causes.
CONSIDER CT SCAN IF FOCAL

DIAZEPAM 5-10 mg I/V or per rectum
OXYGEN (HIGH FLOW, HIGH CONCENTRATION)
PATIENT IN RECOVERY POSITION
PHENYTOIN 18 mg/kg IF NOT LOADED

IS AIRWAY PROTECTED?
SaO₂ >90%?

(YES)

OBSERVE

(NO)

⚠ BEWARE RESPIRATORY DEPRESSION FROM DRUG THERAPY

SEIZURES CONTINUE

DIAZEPAM 5-10 mg BOLUSES
AND / OR
0.8% CHLORMETHIAZOLE 40-100 ml OVER 10 min
TITRATE INFUSION TO CONTROL SEIZURES

INTUBATE + IPPV
(hyperventilate if indicated, e.g. trauma)
CONSIDER BENZODIAZEPINE INFUSION

SEIZURES CONTINUE

SEIZURES CONTINUE

THIOPENTONE INFUSION
INCREASE DOSE FROM 2 mg/kg/hr
+ EEG MONITORING IF POSSIBLE

⚠ THIOPENTONE CAUSES VASODILATATION
± MYOCARDIAL DEPRESSION - MAY NEED FLUID/INOTROPE/PRESSOR THERAPY.
NB THERE IS A RISK OF PANCREATITIS WITH PROLONGED USAGE

SEIZURES CONTINUE

CONSIDER:
• 0.8% CHLORMETHIAZOLE INFUSION
   (if not given already)
• CLONAZEPAM INFUSION 1 mg/hr
• ETOMIDATE INFUSION 25 µg/kg/min
   (steroid cover required)
• MgSO₄

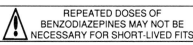

⚠ REPEATED DOSES OF BENZODIAZEPINES MAY NOT BE NECESSARY FOR SHORT-LIVED FITS

⚠ AVOID MUSCLE RELAXATION IF POSSIBLE

7.3/7.4: Brain stem death: potential organ donor

1. A checklist of brainstem death tests and a special chart to record tests is useful.
2. Discussion of brain stem death tests and (if indicated) organ donation must include the family and all the medical and nursing staff.
3. Treatment of many patients with chronic renal failure, cardiac failure, hepatic failure and some forms of chronic respiratory failure are critically dependent on organ donation. Organ donors mainly come from the ICU.
4. Always consider the possibility of the patient becoming an organ donor which, in itself, may be a reason for admission to ICU.
5. Consult the local transplant coordinator early.
6. Do not reject brain dead potential donors e.g. with fully treated infections or acute renal failure without consultation with the transplant coordinator.
7. Maintain organs of brain stem dead patients in optimal condition for transplantation, e.g. avoid hypotension, fluid imbalance, etc. . .
8. A member of the transplant team should not ideally perform brain stem testing.

Bibliography

Cadaveric organs for transplantation. A code of practice including the diagnosis of brain death. Department of Health, London 1983.

Gore SM, Hinds CH, and Rutherford AJ. Organ donation from intensive care units in England. Br Med J 1989; 249: 1193–1197.

Jennett B. Brain death. Br J Anaesth 1981; 53: 111–1119.

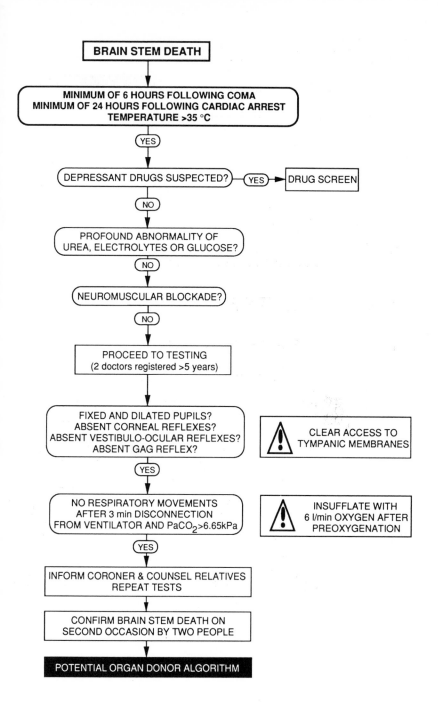

BRAIN STEM DEATH

MINIMUM OF 6 HOURS FOLLOWING COMA
MINIMUM OF 24 HOURS FOLLOWING CARDIAC ARREST
TEMPERATURE >35 °C

YES

DEPRESSANT DRUGS SUSPECTED? — YES → DRUG SCREEN

NO

PROFOUND ABNORMALITY OF
UREA, ELECTROLYTES OR GLUCOSE?

NO

NEUROMUSCULAR BLOCKADE?

NO

PROCEED TO TESTING
(2 doctors registered >5 years)

FIXED AND DILATED PUPILS?
ABSENT CORNEAL REFLEXES?
ABSENT VESTIBULO-OCULAR REFLEXES?
ABSENT GAG REFLEX?

⚠ CLEAR ACCESS TO
TYMPANIC MEMBRANES

YES

NO RESPIRATORY MOVEMENTS
AFTER 3 min DISCONNECTION
FROM VENTILATOR AND PaCO$_2$>6.65kPa

⚠ INSUFFLATE WITH
6 l/min OXYGEN AFTER
PREOXYGENATION

YES

INFORM CORONER & COUNSEL RELATIVES
REPEAT TESTS

CONFIRM BRAIN STEM DEATH ON
SECOND OCCASION BY TWO PEOPLE

POTENTIAL ORGAN DONOR ALGORITHM

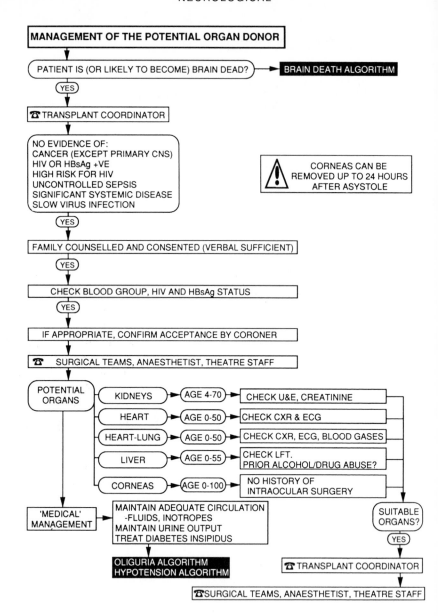

MANAGEMENT OF THE POTENTIAL ORGAN DONOR

PATIENT IS (OR LIKELY TO BECOME) BRAIN DEAD? → **BRAIN DEATH ALGORITHM**

(YES)

☎ TRANSPLANT COORDINATOR

NO EVIDENCE OF:
CANCER (EXCEPT PRIMARY CNS)
HIV OR HBsAg +VE
HIGH RISK FOR HIV
UNCONTROLLED SEPSIS
SIGNIFICANT SYSTEMIC DISEASE
SLOW VIRUS INFECTION

⚠ CORNEAS CAN BE REMOVED UP TO 24 HOURS AFTER ASYSTOLE

(YES)

FAMILY COUNSELLED AND CONSENTED (VERBAL SUFFICIENT)

(YES)

CHECK BLOOD GROUP, HIV AND HBsAg STATUS

(YES)

IF APPROPRIATE, CONFIRM ACCEPTANCE BY CORONER

☎ SURGICAL TEAMS, ANAESTHETIST, THEATRE STAFF

POTENTIAL ORGANS

KIDNEYS → (AGE 4-70) → CHECK U&E, CREATININE

HEART → (AGE 0-50) → CHECK CXR & ECG

HEART-LUNG → (AGE 0-50) → CHECK CXR, ECG, BLOOD GASES

LIVER → (AGE 0-55) → CHECK LFT. PRIOR ALCOHOL/DRUG ABUSE?

CORNEAS → (AGE 0-100) → NO HISTORY OF INTRAOCULAR SURGERY

'MEDICAL' MANAGEMENT

MAINTAIN ADEQUATE CIRCULATION -FLUIDS, INOTROPES
MAINTAIN URINE OUTPUT
TREAT DIABETES INSIPIDUS

**OLIGURIA ALGORITHM
HYPOTENSION ALGORITHM**

SUITABLE ORGANS?

(YES)

☎ TRANSPLANT COORDINATOR

☎ SURGICAL TEAMS, ANAESTHETIST, THEATRE STAFF

Notes

8. Multisystem disease

8.1: Sepsis syndrome

1. This syndrome is an uncontrolled host inflammatory response to infection or tissue damage.
2. Control of the sepsis syndrome requires early and aggressive removal of the source.
   - Drainage of pus.
   - Broad spectrum parenteral antibiotic therapy.
   - Aggressive debridement of all unclean wounds.
   - Early fixation of fractures.

Physiological defects in the sepsis syndrome
1. Septicaemic shock is associated with failure of all components of the circulation.
   - Relative hypovolaemia.
   - Cardiac dysfunction.
   - Peripheral vascular failure.
2. Peripheral vascular failure and right ventricular failure are significant factors in mortality.
3. Normal, resting values of oxygen transport (DO_2) may be inadequate for the high metabolic rate associated with sepsis syndrome.
4. Depressed oxygen extraction reduces tissue oxygen consumption (VO_2) despite normal DO_2.
5. Organ failure is associated with anaerobic metabolism and lactate production.
6. Lactaemia may also be due to impaired hepatic clearance.

Therapeutic goals
1. Achieving supranormal values of cardiac index (>4.5 l/min/m^2), DO_2 (>600 ml/min/m^2) and VO_2 (>170 ml/min/m^2) may improve survival though this awaits large study confirmation.
2. Tissue oxygen debt should be prevented (or removed) by increasing DO_2 until no further rise is seen in VO_2. A satisfactory 'oxygen challenge' should aim to increase DO_2 by 15–20% to rule out methodological errors.
3. Blood lactate levels should fall towards the normal range (<2 mmol/l). However, values within the normal range do not exclude the presence of an oxygen debt.

Volume therapy
1. Due to capillary leak, intravascular volume expansion is often required despite oedema.
2. Stroke volume should usually be maximized with colloid before starting inotropes.
3. Increases in DO_2 achieved by fluid loading increases VO_2 when lactic acidosis or a low cardiac output is present.
4. The stroke volume response to increases in filling pressure is depressed in septic shock.

Inotrope and vasopressor therapy
1. Catecholamine resistance requires high dosages of inotropes and vasopressors.

2. Vasopressors may increase filling pressures thus masking hypovolaemia.
3. Vasopressors may reduce cardiac output and therefore DO_2.
4. If cardiac output requires inotropic support this should be started before using vasopressors.
5. The smallest dose of vasopressor should be used to obtain an acceptable mean arterial pressure.

Vasoactive prostaglandins

1. Reduction of high pulmonary vascular resistance improves RV function and may prevent RV failure.
2. Vasoactive prostaglandins improve gas exchange.
3. Removal of pulmonary capillary occlusion may improve left ventricular filling.

Bibliography

Bone RC, Fisher CJ, Clemmer TP, *et al*. The sepsis syndrome: a valid clinical entity. Crit Care Med 1989; 17: 389–393.

Edwards JD, Brown CS, Nightingale P, Slater R, and Farragher EB. Use of survivors' cardiorespiratory values as therapeutic goals in septic shock. Crit Care Med 1989; 17: 1098–1103.

Ellman H. Capillary permeability in septic patients. Crit Care Med 1984; 12: 629–633.

Groenveld ABJ, Bronsveld W, and Thijs LG. Hemodynamic determinants of mortality in human septic shock. Surgery 1986; 99: 140–152.

Holcroft JW, Vassar M, and Webber CJ. Prostaglandin E1 and survival in patients with the adult respiratory distress syndrome. Ann Surg 1986; 203: 371–378.

Meadows D, Edwards, JD, Wilkins RG, and Nightingale P. Reversal of intractable septic shock with norepinephrine therapy. Crit Care Med 1988; 16: 663–666.

Ognibene FP, Parker MM, Natanson C, Shelhamer JH, and Parillo JE. Depressed left ventricular performance in response to volume infusion in patients with sepsis and septic shock. Chest 1988; 93: 903–910.

Parker MM, Shelhamer JH, Bacharach SL, *et al*. Profound but reversible myocardial depression in patients with septic shock. Ann Intern Med 1984; 100: 483–490.

Parratt JR. Alterations in vascular reactivity in sepsis and endotoxemia. In: Vincent JL ed. Update in Intensive Care and Emergency Medicine 8. Berlin: Springer-Verlag, 1989: 299–308.

Shoemaker WC, Appel PL, Kram HB, Waxman K, and Lee TS. Prospective trial of supranormal values of survivors as therapeutic goals in high-risk patients. Chest 1988; 94: 1176–1186.

Wolf YG, Cotev S, Perel A, and Manny J. Dependance of oxygen consumption on cardiac output in sepsis. Crit Care Med 1987; 15: 198–203.

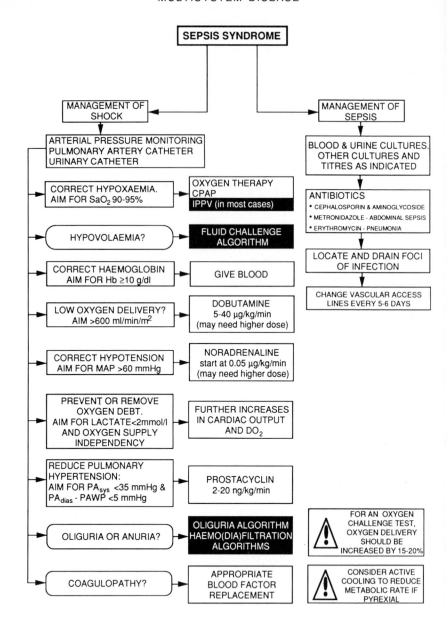

SEPSIS SYNDROME

MANAGEMENT OF SHOCK

ARTERIAL PRESSURE MONITORING
PULMONARY ARTERY CATHETER
URINARY CATHETER

CORRECT HYPOXAEMIA. AIM FOR SaO_2 90-95%	OXYGEN THERAPY CPAP **IPPV (in most cases)**
HYPOVOLAEMIA?	**FLUID CHALLENGE ALGORITHM**
CORRECT HAEMOGLOBIN AIM FOR Hb ≥10 g/dl	GIVE BLOOD
LOW OXYGEN DELIVERY? AIM >600 ml/min/m^2	DOBUTAMINE 5-40 µg/kg/min (may need higher dose)
CORRECT HYPOTENSION AIM FOR MAP >60 mmHg	NORADRENALINE start at 0.05 µg/kg/min (may need higher dose)
PREVENT OR REMOVE OXYGEN DEBT. AIM FOR LACTATE<2mmol/l AND OXYGEN SUPPLY INDEPENDENCY	FURTHER INCREASES IN CARDIAC OUTPUT AND DO_2
REDUCE PULMONARY HYPERTENSION: AIM FOR PA_{sys} <35 mmHg & PA_{dias} - PAWP <5 mmHg	PROSTACYCLIN 2-20 ng/kg/min
OLIGURIA OR ANURIA?	**OLIGURIA ALGORITHM HAEMO(DIA)FILTRATION ALGORITHMS**
COAGULOPATHY?	APPROPRIATE BLOOD FACTOR REPLACEMENT

MANAGEMENT OF SEPSIS

BLOOD & URINE CULTURES. OTHER CULTURES AND TITRES AS INDICATED

ANTIBIOTICS
• CEPHALOSPORIN & AMINOGLYCOSIDE
• METRONIDAZOLE - ABDOMINAL SEPSIS
• ERYTHROMYCIN - PNEUMONIA

LOCATE AND DRAIN FOCI OF INFECTION

CHANGE VASCULAR ACCESS LINES EVERY 5-6 DAYS

⚠ FOR AN OXYGEN CHALLENGE TEST, OXYGEN DELIVERY SHOULD BE INCREASED BY 15-20%

⚠ CONSIDER ACTIVE COOLING TO REDUCE METABOLIC RATE IF PYREXIAL

Notes

8.2: Multiple trauma

History
1. It is crucial to obtain a good history of the event, e.g. time, distance of fall, speed of vehicle, damage to vehicle, death of other vehicular occupants, type of gun and distance from gunman, conscious level and vital signs when found, etc. . .

Neck management
1. If cervical injury suspected the neck must be immediately immobilized (semi-rigid collar, sandbags, etc. . .) and remain so until cleared by an expert.
2. The neck should be manually immobilised during intubation procedures.
3. Normal cervical spine X-rays do not necessarily exclude cord damage.

Respiratory management
1. Ventilation prior to development of respiratory failure may help prevent ARDS.
2. Upper airway obstruction with inability to intubate may require urgent needle cricothyroidotomy followed by surgical cricothyroidotomy, inserting (in adults) a cuffed tracheostomy tube (e.g. 6.0 or greater).

Shock management
1. Restoration of an adequate circulation of oxygenated blood and organ perfusion is crucial.
2. Failure to restore gastrointestinal perfusion in the 'golden hour' immediately following trauma may lead to later organ failure.
3. Volume requirements are commonly underestimated.
4. Cardiac arrest with thoracic trauma requires urgent consideration of:
   - tension pneumothorax
   - massive haemothorax
   - diaphragmatic rupture
   - pericardial tamponade
   - aortic dissection
   - cardiac injury.

   Urgent thoracotomy may be necessary to:
   - facilitate pericardial drainage
   - allow internal massage (particularly with flail chest)
   - to allow aortic cross-clamping to maintain cerebral perfusion (particularly in severe abdominal haemorrhage).
5. Haemorrhage may become more obvious as blood pressure rises.
6. Consider covert haemorrhage, e.g. retroperitoneal, pelvic. Large volumes may also be lost in the thigh, buttock, etc. . .
7. Shock is nearly always due to haemorrhage in the early phase though other causes such as pneumothorax, tamponade and cardiac injury should be considered.
8. A MAST (medical anti-shock trouser) suit, otherwise known as a Pneumatic Anti-shock Garment, raises systemic pressure by increasing venous return, peripheral vascular resistance and left ventricular afterload. In-hospital indications for MAST suit application include:
   - splinting and control of pelvic fractures with continuing haemorrhage and hypotension.
   - for transfer of patients with intra-abdominal trauma and severe hypovolaemia.

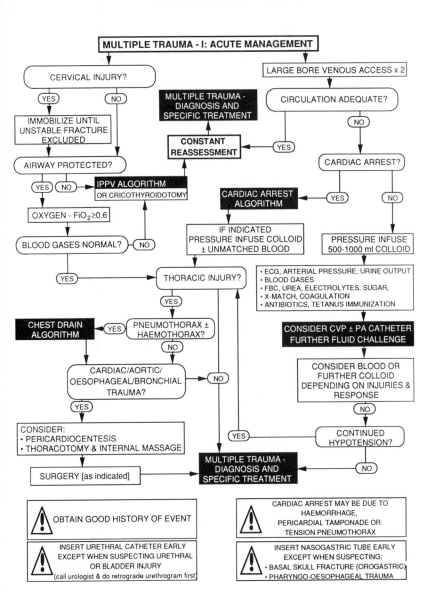

MULTIPLE TRAUMA - I: ACUTE MANAGEMENT

CERVICAL INJURY? — YES / NO

LARGE BORE VENOUS ACCESS x 2

MULTIPLE TRAUMA - DIAGNOSIS AND SPECIFIC TREATMENT

CIRCULATION ADEQUATE? — NO

IMMOBILIZE UNTIL UNSTABLE FRACTURE EXCLUDED

CONSTANT REASSESSMENT — YES

CARDIAC ARREST?

AIRWAY PROTECTED? — YES / NO

IPPV ALGORITHM OR CRICOTHYROIDOTOMY

CARDIAC ARREST ALGORITHM — YES / NO

OXYGEN - FiO$_2$ ≥0.6

IF INDICATED PRESSURE INFUSE COLLOID ± UNMATCHED BLOOD

PRESSURE INFUSE 500-1000 ml COLLOID

BLOOD GASES NORMAL? — NO

THORACIC INJURY? — YES

• ECG, ARTERIAL PRESSURE, URINE OUTPUT
• BLOOD GASES
• FBC, UREA, ELECTROLYTES, SUGAR,
• X-MATCH, COAGULATION
• ANTIBIOTICS, TETANUS IMMUNIZATION

CHEST DRAIN ALGORITHM — YES — PNEUMOTHORAX ± HAEMOTHORAX? — NO

CONSIDER CVP ± PA CATHETER FURTHER FLUID CHALLENGE

CARDIAC/AORTIC/ OESOPHAGEAL/BRONCHIAL TRAUMA? — NO — YES

CONSIDER BLOOD OR FURTHER COLLOID DEPENDING ON INJURIES & RESPONSE — NO

CONSIDER:
• PERICARDIOCENTESIS
• THORACOTOMY & INTERNAL MASSAGE

CONTINUED HYPOTENSION? — YES

SURGERY [as indicated]

MULTIPLE TRAUMA - DIAGNOSIS AND SPECIFIC TREATMENT — NO

⚠ OBTAIN GOOD HISTORY OF EVENT

⚠ CARDIAC ARREST MAY BE DUE TO HAEMORRHAGE, PERICARDIAL TAMPONADE OR TENSION PNEUMOTHORAX

⚠ INSERT URETHRAL CATHETER EARLY EXCEPT WHEN SUSPECTING URETHRAL OR BLADDER INJURY (call urologist & do retrograde urethrogram first)

⚠ INSERT NASOGASTRIC TUBE EARLY EXCEPT WHEN SUSPECTING:
• BASAL SKULL FRACTURE (OROGASTRIC)
• PHARYNGO-OESOPHAGEAL TRAUMA

Contraindications:
- (relative) uncontrolled haemorrhage outside the confines of the garment.
- (absolute) left ventricular dysfunction, diaphragmatic rupture.

Miscellaneous

1. A nasogastric tube and urethral catheter should be inserted early unless contraindicated (see algorithm).
2. Rhabdomyolysis is a common cause of post-traumatic acute renal failure. It should be sought early (creatine kinase levels, myoglobinuria) and treated aggressively (forced diuresis of at least 8–10 l/day for 2–3 days, alkalinizing if necessary to keep urine pH $\geqslant 5$).
3. Likewise compartment syndrome should be considered and treated early. Compartment pressure monitoring is helpful.
4. Steroids have been shown to be efficacious for spinal injury if given within eight hours (methylprednisolone 30 mg/kg bolus then 5.4 mg/kg/hr for 23 hr).
5. Early fixation of pelvic and femoral fractures:
   - facilitates nursing
   - reduces mortality from sepsis and ARDS.
6. Excessive crystalloid fluids should be avoided.

Bibliography

Advanced Trauma Life Support Student Manual. American College of Surgeons, Illinois 1989.

Border JR, Hassett J, LaDuca J, et al. The gut origin of septic states in blunt multiple trauma (ISS$\geqslant$40) in the ICU. Ann Surg 1987; 206: 427–428.

Bracken MB, Shepard MJ, Collins WF et al. A randomised, controlled trial of methylprednisolone or naloxone in the treatment of acute spinal cord injury. Results of the Second National Acute Spinal Cord Injury Study. N Engl J Med 1990; 322: 1405–1411.

Fiddian-Green RD. Splanchnic ischaemia and multiple organ failure in the critically ill. Ann R Coll Surg Engl 1988; 70: 128–134.

Johnson KD, Cadambi A, and Seibert GB. Incidence of adult respiratory distress syndrome in patients with multiple musculoskeletal injuries; effect of early operative stabilisation of fractures. J Trauma 1985; 25: 375–384.

Porter K. Trauma care—clinical assessment and management. Care of the Critically Ill 1987; 3: 180–183.

Richardson JD, Franz JL, Grover FL, and Trinkle JK. Pulmonary contusion and haemorrhage—crystalloid versus colloid replacement. J Surg Res 1974; 16: 330–336.

Rush BF, Sori AJ, Murphy TF, Smith S, Flanagan JJ, and Machiedo GW. Endotoxemia and bacteremia during hemorrhagic shock. Ann Surg 1988; 207: 549–554.

van Niekerk J and Goris RJA. Management of the trauma patient. Clin Intensive Care 1990; 1: 32–36.

Wilder-Baker J, Deitch EA, Berg RD, and Specian RD. Hemorrhagic shock induces bacterial translocation from the gut. J. Trauma 1988; 28: 896–906.

MULTIPLE TRAUMA - II: DIAGNOSIS AND SPECIFIC TREATMENT

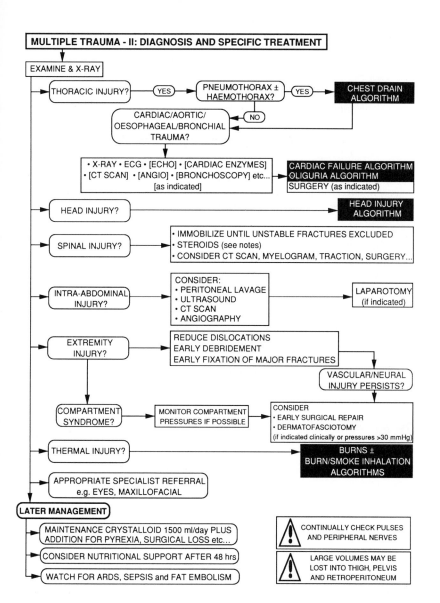

EXAMINE & X-RAY

THORACIC INJURY? — YES → PNEUMOTHORAX ± HAEMOTHORAX? — YES → **CHEST DRAIN ALGORITHM**

NO

CARDIAC/AORTIC/ OESOPHAGEAL/BRONCHIAL TRAUMA?

• X-RAY • ECG • [ECHO] • [CARDIAC ENZYMES] • [CT SCAN] • [ANGIO] • [BRONCHOSCOPY] etc... [as indicated]

CARDIAC FAILURE ALGORITHM OLIGURIA ALGORITHM SURGERY (as indicated)

HEAD INJURY? → **HEAD INJURY ALGORITHM**

SPINAL INJURY? →
• IMMOBILIZE UNTIL UNSTABLE FRACTURES EXCLUDED
• STEROIDS (see notes)
• CONSIDER CT SCAN, MYELOGRAM, TRACTION, SURGERY...

INTRA-ABDOMINAL INJURY? →
CONSIDER:
• PERITONEAL LAVAGE
• ULTRASOUND
• CT SCAN
• ANGIOGRAPHY
→ LAPAROTOMY (if indicated)

EXTREMITY INJURY? →
REDUCE DISLOCATIONS
EARLY DEBRIDEMENT
EARLY FIXATION OF MAJOR FRACTURES
→ VASCULAR/NEURAL INJURY PERSISTS?

COMPARTMENT SYNDROME? → MONITOR COMPARTMENT PRESSURES IF POSSIBLE →
CONSIDER
• EARLY SURGICAL REPAIR
• DERMATOFASCIOTOMY
(if indicated clinically or pressures >30 mmHg)

THERMAL INJURY? → **BURNS ± BURN/SMOKE INHALATION ALGORITHMS**

APPROPRIATE SPECIALIST REFERRAL e.g. EYES, MAXILLOFACIAL

LATER MANAGEMENT

MAINTENANCE CRYSTALLOID 1500 ml/day PLUS ADDITION FOR PYREXIA, SURGICAL LOSS etc...

CONSIDER NUTRITIONAL SUPPORT AFTER 48 hrs

WATCH FOR ARDS, SEPSIS and FAT EMBOLISM

⚠ CONTINUALLY CHECK PULSES AND PERIPHERAL NERVES

⚠ LARGE VOLUMES MAY BE LOST INTO THIGH, PELVIS AND RETROPERITONEUM

8.3: Burns management

Days 1–2
1. Colloid or crystalloid solutions can be used provided sufficient volume and sodium is given.
2. Colloids may reduce oedema at non-burn sites and restore blood volume more quickly than crystalloids. If crystalloids alone are used, requirements are 4 ml/kg/%burn/24 hr compared to colloid 2.5 ml/kg/%burn/24 hr (q.v.).
3. Sodium needs are 0.5 mmol/kg/%burn/24 hr.
4. Evaporative losses are normally 2 ml/kg/hr. Water losses may be increased if wounds are not covered.
5. Avoid overzealous fluid infusion so that oedema is minimized.
6. Formulae are to be used as guidelines only. Constant reassessment is essential.
7. The Parkland (U.S.A.) formula divides the first 24 hrs into 3 8-hr periods.
   2 ml/kg/%burn of Hartmann's solution is given in the first period.
   1 ml/kg/%burn of Hartmann's solution is given in the second period.
   1 ml/kg/%burn of Hartmann's solution is given in the third period.
   In the second 24-hr period 0.3–0.5 ml plasma/kg/%burn is given.

Days 2–5
1. Oedema reabsorbs and Na^+ moves out of cells. Fluid and Na^+ replacement should therefore be reduced.
2. Potassium needs will increase as urinary output rises.
3. Commence nutrition. Increase gradually to:
   - Protein 1 g/kg + 2 g/ % burn.
   - Calories 20 Cal/kg + 50 Cal/ % burn.

Day 6—wound closure
1. There is a marked increase in metabolic rate, temperature, VO_2 and CO_2 production.
2. Patients on mechanical ventilation may require an increased minute volume.
3. Because of 'burn fever' sepsis is difficult to detect. Note increasing base deficit.

Renal failure
1. Early recognition is important.
2. Signs include: urine output <0.5 ml/kg/hr, 'fixed' urine osmolality of approx. 350 mOsm/l, rising creatinine, urine:plasma urea ratio <5, hyperkalaemia and acidosis.

Problems associated with anaesthesia
1. Early grafting:
   - Blood loss may be rapid and massive (prepare to lose 100 ml per 1% of body surface to be grafted).
   - Heat and fluid loss.
   - Hypovolaemia.
2. Late grafting:
   - Contractures may make intubation difficult.
   - Suxamethonium to be avoided at 5–150 days post/burn.
   - Resistance to non-depolarising relaxants.
   - Patients toxic, wasted, anaemic. Often hyponatraemic and hypokalaemic.

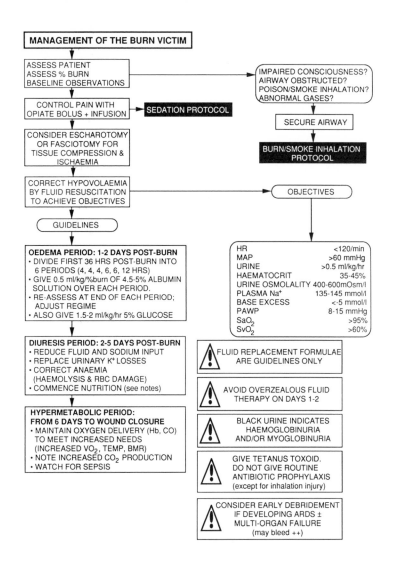

MANAGEMENT OF THE BURN VICTIM

ASSESS PATIENT
ASSESS % BURN
BASELINE OBSERVATIONS

IMPAIRED CONSCIOUSNESS?
AIRWAY OBSTRUCTED?
POISON/SMOKE INHALATION?
ABNORMAL GASES?

CONTROL PAIN WITH
OPIATE BOLUS + INFUSION → SEDATION PROTOCOL

SECURE AIRWAY

CONSIDER ESCHAROTOMY
OR FASCIOTOMY FOR
TISSUE COMPRESSION &
ISCHAEMIA

BURN/SMOKE INHALATION
PROTOCOL

CORRECT HYPOVOLAEMIA
BY FLUID RESUSCITATION
TO ACHIEVE OBJECTIVES

OBJECTIVES

GUIDELINES

OEDEMA PERIOD: 1-2 DAYS POST-BURN
• DIVIDE FIRST 36 HRS POST-BURN INTO
  6 PERIODS (4, 4, 4, 6, 6, 12 HRS)
• GIVE 0.5 ml/kg/%burn OF 4.5-5% ALBUMIN
  SOLUTION OVER EACH PERIOD.
• RE-ASSESS AT END OF EACH PERIOD;
  ADJUST REGIME
• ALSO GIVE 1.5-2 ml/kg/hr 5% GLUCOSE

HR	<120/min
MAP	>60 mmHg
URINE	>0.5 ml/kg/hr
HAEMATOCRIT	35-45%
URINE OSMOLALITY	400-600mOsm/l
PLASMA Na^+	135-145 mmol/l
BASE EXCESS	<-5 mmol/l
PAWP	8-15 mmHg
SaO_2	>95%
SvO_2	>60%

DIURESIS PERIOD: 2-5 DAYS POST-BURN
• REDUCE FLUID AND SODIUM INPUT
• REPLACE URINARY K^+ LOSSES
• CORRECT ANAEMIA
  (HAEMOLYSIS & RBC DAMAGE)
• COMMENCE NUTRITION (see notes)

⚠ FLUID REPLACEMENT FORMULAE
ARE GUIDELINES ONLY

⚠ AVOID OVERZEALOUS FLUID
THERAPY ON DAYS 1-2

HYPERMETABOLIC PERIOD:
FROM 6 DAYS TO WOUND CLOSURE
• MAINTAIN OXYGEN DELIVERY (Hb, CO)
  TO MEET INCREASED NEEDS
  (INCREASED VO_2, TEMP, BMR)
• NOTE INCREASED CO_2 PRODUCTION
• WATCH FOR SEPSIS

⚠ BLACK URINE INDICATES
HAEMOGLOBINURIA
AND/OR MYOGLOBINURIA

⚠ GIVE TETANUS TOXOID.
DO NOT GIVE ROUTINE
ANTIBIOTIC PROPHYLAXIS
(except for inhalation injury)

⚠ CONSIDER EARLY DEBRIDEMENT
IF DEVELOPING ARDS ±
MULTI-ORGAN FAILURE
(may bleed ++)

Bibliography

Burns and Plastic Surgery. Ed: Judkins KC. Ballieres Clinical Anaesthesiology 1987;
1(3).

Carjaval HF. Optimal composition of burn resuscitation fluid. Crit Care Med 1988; 16:
695–700.

Demling RH. Fluid resuscitation after major burns. JAMA 1983; 250: 1438–1442.

Demling RH. Burns. N Engl J Med. 1985; 313: 1389–1398.

Martyn J. Clinical pharmacology and drug therapy in the burned patient.
Anesthesiology 1986; 65: 65–75.

Pruitt BA. Fluid resuscitation of extensively burned patients. J Trauma 1981; 21
(Suppl.8): 690–692.

Robertson C and Fenton O. ABC of Major Trauma: Management of severe burns. Br
Med J 1990; 301: 282–286.

Settle JAD. Fluid therapy in burns. J Roy Soc Med. 1982; 75 (Suppl.1) 6–11.

Scheulen J and Munster A. The Parkland formula in patients with burns and
inhalation injury. J Trauma 1982; 22: 869–871.

Sutherland AB. Nitrogen balance and nutritional requirement in the burn patient: a
reappraisal. Burns 1976; 2: 238–244.

Appendix

Adrenaline / Noradrenaline / Isoprenaline 2 mg in 50 ml

Infusion volumes given as ml/hr

Dose (µg/kg/min)	Weight (kg)																		
	30	35	40	45	50	55	60	65	70	75	80	85	90	95	100	105	110	115	120
0.01	0.45	0.53	0.6	0.68	0.75	0.83	0.9	0.98	1.05	1.13	1.2	1.28	1.35	1.43	1.5	1.58	1.65	1.73	1.8
0.02	0.9	1.05	1.2	1.35	1.5	1.65	1.8	1.95	2.1	2.25	2.4	2.55	2.7	2.85	3	3.15	3.3	3.45	3.6
0.03	1.35	1.57	1.8	2.02	2.25	2.48	2.7	2.92	3.15	3.38	3.6	3.82	4.05	4.28	4.5	4.72	4.95	5.17	5.4
0.04	1.8	2.1	2.4	2.7	3	3.3	3.6	3.9	4.2	4.5	4.8	5.1	5.4	5.7	6	6.3	6.6	6.9	7.2
0.05	2.25	2.63	3	3.38	3.75	4.13	4.5	4.88	5.25	5.63	6	6.38	6.75	7.13	7.5	7.88	8.25	8.63	9
0.06	2.7	3.15	3.6	4.05	4.5	4.95	5.4	5.85	6.3	6.75	7.2	7.65	8.1	8.55	9	9.45	9.9	10.4	10.8
0.07	3.15	3.68	4.2	4.73	5.25	5.78	6.3	6.83	7.35	7.88	8.4	8.93	9.45	9.98	10.5	11	11.6	12.1	12.6
0.08	3.6	4.2	4.8	5.4	6	6.6	7.2	7.8	8.4	9	9.6	10.2	10.8	11.4	12	12.6	13.2	13.8	14.4
0.09	4.05	4.72	5.4	6.08	6.75	7.42	8.1	8.78	9.45	10.1	10.8	11.5	12.2	12.8	13.5	14.2	14.9	15.5	16.2
0.1	4.5	5.25	6	6.75	7.5	8.25	9	9.75	10.5	11.3	12	12.8	13.5	14.3	15	15.8	16.5	17.3	18
0.2	9	10.5	12	13.5	15	16.5	18	19.5	21	22.5	24	25.5	27	28.5	30	31.5	33	34.5	36
0.3	13.5	15.8	18	20.3	22.5	24.8	27	29.3	31.5	33.8	36	38.3	40.5	42.8	45	47.3	49.5	51.8	54
0.4	18	21	24	27	30	33	36	39	42	45	48	51	54	57	60	63	66	69	72
0.5	22.5	26.3	30	33.8	37.5	41.3	45	48.8	52.5	56.3	60	63.8	67.5	71.3	75	78.8	82.5	86.3	90
0.6	27	31.5	36	40.5	45	49.5	54	58.5	63	67.5	72	76.5	81	85.5	90	94.5	99	104	108
0.7	31.5	36.8	42	47.3	52.5	57.8	63	68.3	73.5	78.8	84	89.3	94.5	99.8	105	110	116	121	126
0.8	36	42	48	54	60	66	72	78	84	90	96	102	108	114	120	126	132	138	144
0.9	40.5	47.3	54	60.8	67.5	74.3	81	87.8	94.5	101	108	115	122	128	135	142	149	155	162
1	45	52.5	60	67.5	75	82.5	90	97.5	105	113	120	128	135	143	150	158	165	173	180

Adrenaline / Noradrenaline / Isoprenaline 4 mg in 50 ml

Infusion volumes given as ml/hr

Dose (µg/kg/min)	Weight (kg)																		
	30	35	40	45	50	55	60	65	70	75	80	85	90	95	100	105	110	115	120
0.01	0.23	0.26	0.3	0.34	0.38	0.41	0.45	0.49	0.53	0.56	0.6	0.64	0.68	0.71	0.75	0.79	0.83	0.86	0.9
0.02	0.45	0.53	0.6	0.68	0.75	0.83	0.9	0.98	1.05	1.13	1.2	1.28	1.35	1.43	1.5	1.58	1.65	1.73	1.8
0.03	0.68	0.79	0.9	1.01	1.13	1.24	1.35	1.46	1.57	1.69	1.8	1.91	2.02	2.14	2.25	2.36	2.48	2.59	2.7
0.04	0.9	1.05	1.2	1.35	1.5	1.65	1.8	1.95	2.1	2.25	2.4	2.55	2.7	2.85	3	3.15	3.3	3.45	3.6
0.05	1.13	1.31	1.5	1.69	1.88	2.06	2.25	2.44	2.63	2.81	3	3.19	3.38	3.56	3.75	3.94	4.13	4.31	4.5
0.06	1.35	1.57	1.8	2.02	2.25	2.48	2.7	2.92	3.15	3.38	3.6	3.82	4.05	4.28	4.5	4.72	4.95	5.17	5.4
0.07	1.58	1.84	2.1	2.36	2.63	2.89	3.15	3.41	3.68	3.94	4.2	4.46	4.73	4.99	5.25	5.51	5.78	6.04	6.3
0.08	1.8	2.1	2.4	2.7	3	3.3	3.6	3.9	4.2	4.5	4.8	5.1	5.4	5.7	6	6.3	6.6	6.9	7.2
0.09	2.02	2.36	2.7	3.04	3.38	3.71	4.05	4.39	4.72	5.06	5.4	5.74	6.08	6.41	6.75	7.09	7.42	7.76	8.1
0.1	2.25	2.63	3	3.38	3.75	4.13	4.5	4.88	5.25	5.63	6	6.38	6.75	7.13	7.5	7.88	8.25	8.63	9
0.2	4.5	5.25	6	6.75	7.5	8.25	9	9.75	10.5	11.3	12	12.8	13.5	14.3	15	15.8	16.5	17.3	18
0.3	6.75	7.88	9	10.1	11.3	12.4	13.5	14.6	15.8	16.9	18	19.1	20.3	21.4	22.5	23.6	24.8	25.9	27
0.4	9	10.5	12	13.5	15	16.5	18	19.5	21	22.5	24	25.5	27	28.5	30	31.5	33	34.5	36
0.5	11.3	13.1	15	16.9	18.8	20.6	22.5	24.4	26.3	28.1	30	31.9	33.8	35.6	37.5	39.4	41.3	43.1	45
0.6	13.5	15.8	18	20.3	22.5	24.8	27	29.3	31.5	33.8	36	38.3	40.5	42.8	45	47.3	49.5	51.8	54
0.7	15.8	18.4	21	23.6	26.3	28.9	31.5	34.1	36.8	39.4	42	44.6	47.3	49.9	52.5	55.1	57.8	60.4	63
0.8	18	21	24	27	30	33	36	39	42	45	48	51	54	57	60	63	66	69	72
0.9	20.3	23.6	27	30.4	33.8	37.1	40.5	43.9	47.3	50.6	54	57.4	60.8	64.1	67.5	70.9	74.3	77.6	81
1	22.5	26.3	30	33.8	37.5	41.3	45	48.8	52.5	56.3	60	63.8	67.5	71.3	75	78.8	82.5	86.3	90
2	45	52.5	60	67.5	75	82.5	90	97.5	105	113	120	128	135	143	150	158	165	173	180
3	67.5	78.8	90	101	113	124	135	146	158	169	180	191	203	214	225	236	248	259	270
4	90	105	120	135	150	165	180	195	210	225	240	255	270	285	300	315	330	345	360

Adrenaline / Noradrenaline / Isoprenaline 8 mg in 50 ml

Infusion volumes given as ml/hr

Weight (kg)

Dose (µg/kg/min)	30	35	40	45	50	55	60	65	70	75	80	85	90	95	100	105	110	115	120
0.1	1.13	1.31	1.5	1.69	1.88	2.06	2.25	2.44	2.63	2.81	3	3.19	3.38	3.56	3.75	3.94	4.13	4.31	4.5
0.2	2.25	2.63	3	3.38	3.75	4.13	4.5	4.88	5.25	5.63	6	6.38	6.75	7.13	7.5	7.88	8.25	8.63	9
0.3	3.38	3.94	4.5	5.06	5.63	6.19	6.75	7.31	7.88	8.44	9	9.56	10.1	10.7	11.3	11.8	12.4	12.9	13.5
0.4	4.5	5.25	6	6.75	7.5	8.25	9	9.75	10.5	11.3	12	12.8	13.5	14.3	15	15.8	16.5	17.3	18
0.5	5.63	6.56	7.5	8.44	9.38	10.3	11.3	12.2	13.1	14.1	15	15.9	16.9	17.8	18.8	19.7	20.6	21.6	22.5
0.6	6.75	7.88	9	10.1	11.3	12.4	13.5	14.6	15.8	16.9	18	19.1	20.3	21.4	22.5	23.6	24.8	25.9	27
0.7	7.88	9.19	10.5	11.8	13.1	14.4	15.8	17.1	18.4	19.7	21	22.3	23.6	24.9	26.3	27.6	28.9	30.2	31.5
0.8	9	10.5	12	13.5	15	16.5	18	19.5	21	22.5	24	25.5	27	28.5	30	31.5	33	34.5	36
0.9	10.1	11.8	13.5	15.2	16.9	18.6	20.3	21.9	23.6	25.3	27	28.7	30.4	32.1	33.8	35.4	37.1	38.8	40.5
1	11.3	13.1	15	16.9	18.8	20.6	22.5	24.4	26.3	28.1	30	31.9	33.8	35.6	37.5	39.4	41.3	43.1	45
2	22.5	26.3	30	33.8	37.5	41.3	45	48.8	52.5	56.3	60	63.8	67.5	71.3	75	78.8	82.5	86.3	90
3	33.8	39.4	45	50.6	56.3	61.9	67.5	73.1	78.8	84.4	90	95.6	101	107	113	118	124	129	135
4	45	52.5	60	67.5	75	82.5	90	97.5	105	113	120	128	135	143	150	158	165	173	180
5	56.3	65.6	75	84.4	93.8	103	113	122	131	141	150	159	169	178	188	197	206	216	225
6	67.5	78.8	90	101	113	124	135	146	158	169	180	191	203	214	225	236	248	259	270
7	78.8	91.9	105	118	131	144	158	171	184	197	210	223	236	249	263	276	289	302	315
8	90	105	120	135	150	165	180	195	210	225	240	255	270	285	300	315	330	345	360
9	101	118	135	152	169	186	203	219	236	253	270	287	304	321	338	354	371	388	405
10	113	131	150	169	188	206	225	244	263	281	300	319	338	356	375	394	413	431	450

Dobutamine 250 mg in 50 ml

Infusion volume given as ml/hr

Weight (kg)

Dose (µg/mg/min)	30	35	40	45	50	55	60	65	70	75	80	85	90	95	100	105	110	115	120
1	0.36	0.42	0.48	0.54	0.6	0.66	0.72	0.78	0.84	0.9	0.96	1.02	1.08	1.14	1.2	1.26	1.32	1.38	1.44
2	0.72	0.84	0.96	1.08	1.2	1.32	1.44	1.56	1.68	1.8	1.92	2.04	2.16	2.28	2.4	2.52	2.64	2.76	2.88
3	1.08	1.26	1.44	1.62	1.8	1.98	2.16	2.34	2.52	2.7	2.88	3.06	3.24	3.42	3.6	3.78	3.96	4.14	4.32
4	1.44	1.68	1.92	2.16	2.4	2.64	2.88	3.12	3.36	3.6	3.84	4.08	4.32	4.56	4.8	5.04	5.28	5.52	5.76
5	1.8	2.1	2.4	2.7	3	3.3	3.6	3.9	4.2	4.5	4.8	5.1	5.4	5.7	6	6.3	6.6	6.9	7.2
6	2.16	2.52	2.88	3.24	3.6	3.96	4.32	4.68	5.04	5.4	5.76	6.12	6.48	6.84	7.2	7.56	7.92	8.28	8.64
7	2.52	2.94	3.36	3.78	4.2	4.62	5.04	5.46	5.88	6.3	6.72	7.14	7.56	7.98	8.4	8.82	9.24	9.66	10.1
8	2.88	3.36	3.84	4.32	4.8	5.28	5.76	6.24	6.72	7.2	7.68	8.16	8.64	9.12	9.6	10.1	10.6	11	11.5
9	3.24	3.78	4.32	4.86	5.4	5.94	6.48	7.02	7.56	8.1	8.64	9.18	9.72	10.3	10.8	11.3	11.9	12.4	13
10	3.6	4.2	4.8	5.4	6	6.6	7.2	7.8	8.4	9	9.6	10.2	10.8	11.4	12	12.6	13.2	13.8	14.4
20	7.2	8.4	9.6	10.8	12	13.2	14.4	15.6	16.8	18	19.2	20.4	21.6	22.8	24	25.2	26.4	27.6	28.8
30	10.8	12.6	14.4	16.2	18	19.8	21.6	23.4	25.2	27	28.8	30.6	32.4	34.2	36	37.8	39.6	41.4	43.2
40	14.4	16.8	19.2	21.6	24	26.4	28.8	31.2	33.6	36	38.4	40.8	43.2	45.6	48	50.4	52.8	55.2	57.6
50	18	21	24	27	30	33	36	39	42	45	48	51	54	57	60	63	66	69	72
60	21.6	25.2	28.8	32.4	36	39.6	43.2	46.8	50.4	54	57.6	61.2	64.8	68.4	72	75.6	79.2	82.8	86.4
70	25.2	29.4	33.6	37.8	42	46.2	50.4	54.6	58.8	63	67.2	71.4	75.6	79.8	84	88.2	92.4	96.6	101
80	28.8	33.6	38.4	43.2	48	52.8	57.6	62.4	67.2	72	76.8	81.6	86.4	91.2	96	101	106	110	115
90	32.4	37.8	43.2	48.6	54	59.4	64.8	70.2	75.6	81	86.4	91.8	97.2	103	108	113	119	124	130
100	36	42	48	54	60	66	72	78	84	90	96	102	108	114	120	126	132	138	144

Dobutamine 500 mg in 50 ml

Infusion volume given as ml/hr

Dose (µg/kg/min)	Weight (kg)																		
	30	35	40	45	50	55	60	65	70	75	80	85	90	95	100	105	110	115	120
1	0.18	0.21	0.24	0.27	0.3	0.33	0.36	0.39	0.42	0.45	0.48	0.51	0.54	0.57	0.6	0.63	0.66	0.69	0.72
2	0.36	0.42	0.48	0.54	0.6	0.66	0.72	0.78	0.84	0.9	0.96	1.02	1.08	1.14	1.2	1.26	1.32	1.38	1.44
3	0.54	0.63	0.72	0.81	0.9	0.99	1.08	1.17	1.26	1.35	1.44	1.53	1.62	1.71	1.8	1.89	1.98	2.07	2.16
4	0.72	0.84	0.96	1.08	1.2	1.32	1.44	1.56	1.68	1.8	1.92	2.04	2.16	2.28	2.4	2.52	2.64	2.76	2.88
5	0.9	1.05	1.2	1.35	1.5	1.65	1.8	1.95	2.1	2.25	2.4	2.55	2.7	2.85	3	3.15	3.3	3.45	3.6
6	1.08	1.26	1.44	1.62	1.8	1.98	2.16	2.34	2.52	2.7	2.88	3.06	3.24	3.42	3.6	3.78	3.96	4.14	4.32
7	1.26	1.47	1.68	1.89	2.1	2.31	2.52	2.73	2.94	3.15	3.36	3.57	3.78	3.99	4.2	4.41	4.62	4.83	5.04
8	1.44	1.68	1.92	2.16	2.4	2.64	2.88	3.12	3.36	3.6	3.84	4.08	4.32	4.56	4.8	5.04	5.28	5.52	5.76
9	1.62	1.89	2.16	2.43	2.7	2.97	3.24	3.51	3.78	4.05	4.32	4.59	4.86	5.13	5.4	5.67	5.94	6.21	6.48
10	1.8	2.1	2.4	2.7	3	3.3	3.6	3.9	4.2	4.5	4.8	5.1	5.4	5.7	6	6.3	6.6	6.9	7.2
20	3.6	4.2	4.8	5.4	6	6.6	7.2	7.8	8.4	9	9.6	10.2	10.8	11.4	12	12.6	13.2	13.8	14.4
30	5.4	6.3	7.2	8.1	9	9.9	10.8	11.7	12.6	13.5	14.4	15.3	16.2	17.1	18	18.9	19.8	20.7	21.6
40	7.2	8.4	9.6	10.8	12	13.2	14.4	15.6	16.8	18	19.2	20.4	21.6	22.8	24	25.2	26.4	27.6	28.8
50	9	10.5	12	13.5	15	16.5	18	19.5	21	22.5	24	25.5	27	28.5	30	31.5	33	34.5	36
60	10.8	12.6	14.4	16.2	18	19.8	21.6	23.4	25.2	27	28.8	30.6	32.4	34.2	36	37.8	39.6	41.4	43.2
70	12.6	14.7	16.8	18.9	21	23.1	25.2	27.3	29.4	31.5	33.6	35.7	37.8	39.9	42	44.1	46.2	48.3	50.4
80	14.4	16.8	19.2	21.6	24	26.4	28.8	31.2	33.6	36	38.4	40.8	43.2	45.6	48	50.4	52.8	55.2	57.6
90	16.2	18.9	21.6	24.3	27	29.7	32.4	35.1	37.8	40.5	43.2	45.9	48.6	51.3	54	56.7	59.4	62.1	64.8
100	18	21	24	27	30	33	36	39	42	45	48	51	54	57	60	63	66	69	72

Dopamine 2.5 µg/kg/min

Infusion volumes given as ml/hr

Weight (kg)

	30	35	40	45	50	55	60	65	70	75	80	85	90	95	100	105	110	115	120
Dopamine 200 mg in 50 ml	1.1	1.3	1.5	1.7	1.9	2.1	2.3	2.4	2.6	2.8	3	3.2	3.4	3.6	3.8	3.9	4.1	4.3	4.5
Dopamine 400 mg in 50 ml	0.6	0.7	0.8	0.8	0.9	1	1.1	1.2	1.3	1.4	1.5	1.6	1.7	1.8	1.9	2	2.1	2.2	2.3
Dopamine 800 mg in 50 ml	0.3	0.3	0.4	0.4	0.5	0.5	0.6	0.6	0.7	0.7	0.8	0.8	0.8	0.9	0.9	1	1	1.1	1.1

Index

(Individual drugs are given in *italic type*; main references to algorithms in **bold type**.)